186
Advances in Biochemical Engineering/Biotechnology

Aims and Scope

This book series reviews current trends in modern biotechnology and biochemical engineering. Its aim is to cover all aspects of these interdisciplinary disciplines, where knowledge, methods and expertise are required from chemistry, biochemistry, microbiology, molecular biology, chemical engineering and computer science.

Volumes are organized topically and provide a comprehensive discussion of developments in the field over the past 3–5 years. The series also discusses new discoveries and applications. Special volumes are dedicated to selected topics which focus on new biotechnological products and new processes for their synthesis and purification.

In general, volumes are edited by well-known guest editors. The series editor and publisher will, however, always be pleased to receive suggestions and supplementary information. Manuscripts are accepted in English.

In references, Advances in Biochemical Engineering/Biotechnology is abbreviated as *Adv. Biochem. Engin./Biotechnol.* and cited as a journal.

Yuan Lu • Michael C. Jewett
Editors

Cell-free Production

System Development

With contributions by

S. K. Dondapati · L. Fan · D. T. Gonzales · K. Hagino ·
K. Honda · N. Ichihashi · S. Kubick · Q. Li · X. Ning ·
Y. Pandey · A. T. Patterson · Y. Qin · K. Seo ·
M. P. Styczynski · S. Suraritdechachai · G. Suryatin Alim ·
T. Suzuki · T.-Y. D. Tang · X. Wei · H. Wu ·
D. Wüstenhagen · C. You · B. Zheng

Springer

Editors
Yuan Lu
Department of Chemical Engineering
Tsinghua University
Beijing, China

Michael C. Jewett
Department of Bioengineering
Stanford University
Stanford, CA, USA

ISSN 0724-6145 ISSN 1616-8542 (electronic)
Advances in Biochemical Engineering/Biotechnology
ISBN 978-3-031-43024-4 ISBN 978-3-031-43025-1 (eBook)
https://doi.org/10.1007/978-3-031-43025-1

This Springer imprint is published by the registered company Springer Nature Switzerland AG
The registered company address is: Gewerbestrasse 11, 6330 Cham, Switzerland

Paper in this product is recyclable.

This book is dedicated to James "Jim" R. Swartz. Jim is the James H. Clark Professor in the School of Engineering and Professor of Chemical Engineering and of Bioengineering at Stanford University. He is a world leader of the large, vibrant, and international biochemical engineering community and a founder of the cell-free biotechnology field. Jim's fundamental, original, and pioneering contributions to the field, including seminal bioprocess development innovations, enabled the dawn of a new era in cell-free protein synthesis. This led to a new sector in the biotech industry that is transforming production of protein therapeutics, personalized medicines, vaccines, diagnostics, and value-added biochemicals.

Preface

Cell-free synthetic biology is developing as one of the most potent and effective means of understanding, harnessing, and extending the functions of natural biological systems without using whole living cells. Cell-free biosynthesis, also known as in vitro cell-free transcription translation technology, has been a fundamental and applied biology research tool for nearly seven decades. Cell-free synthetic systems were initially developed in the 1960s and played an essential role in discovering the genetic code. Over the past ten years, cell-free systems have rapidly developed to meet the growing demand for fast recombinant protein expression technologies, leading to the development of numerous highly active cell-free biosynthesis platforms.

With the rapid development of cell-free systems, the types have varied. One is the extract system, which can be divided into high-use model cell types and low-use non-model cell types. The cell types with high use rates mainly include *Escherichia coli*, yeast, wheat germ, insect, and Chinese hamster ovary. These systems have their own advantages and disadvantages, and the appropriate extract can be selected according to the type of protein that needs to be expressed. The use of non-mode cells as a chassis for the construction of cell-free systems, although yet to be widely used or developed, is highly likely to become a source of rapid innovation in cell-free systems in the future. Another type is a cell-free system that uses recombinant elements to synthesize proteins. The technique combines a minimum number of transcription and translation elements. Although the cost is higher than the extract system, because the composition is precise and controllable, it has a significant advantage in basic biochemistry research.

Due to the diversity of cell-free systems and their advantages at different levels, their application fields are becoming increasingly extensive. Nowadays, cell-free systems have shown great application potential in gene circuit research, protein engineering, and artificial cell construction. The future development of cell-free synthesis systems requires further exploitation of their flexibility and high-throughput potential for functional protein screening and continuous growth for detecting and treating novel diseases. It further uses its advantages, such as open characteristics and easy integration of biology, materials science, chemistry, physics, computer science, and other disciplines, to play its vast potential in biomaterials, biocatalysis, medicine, and health.

We hope that this collection of chapters with topics regarding the development of cell-free systems will be helpful, not only for experts in the field of cell-free synthetic biology but also for beginners just starting their academic careers. These chapters cover important current aspects, including system transformation, materials-integrated systems, and cutting-edge applications.

We would like to express our heartfelt gratitude to all the authors and editors for their valuable contributions. Your hard work and dedication have truly made a difference. Thank you for sharing your knowledge and expertise with us. Your contributions have enriched our project and we are truly honored by your participation.

Beijing, China — Yuan Lu
Stanford, CA, USA — Michael C. Jewett

Contents

Biomanufacturing by In Vitro Biotransformation (ivBT) Using Purified Cascade Multi-enzymes 1
Yanmei Qin, Qiangzi Li, Lin Fan, Xiao Ning, Xinlei Wei, and Chun You

Cell-Free Production and Regeneration of Cofactors 29
Gladwin Suryatin Alim, Takuma Suzuki, and Kohsuke Honda

Hydrogel-Based Multi-enzymatic System for Biosynthesis 51
Han Wu and Bo Zheng

Compartmentalized Cell-Free Expression Systems for Building Synthetic Cells 77
David T. Gonzales, Surased Suraritdechachai, and T. -Y. Dora Tang

Cell-Free Synthesis and Electrophysiological Analysis of Multipass Voltage-Gated Ion Channels Tethered in Microsomal Membranes 103
Yogesh Pandey, Srujan Kumar Dondapati, Doreen Wüstenhagen, and Stefan Kubick

Progresses in Cell-Free In Vitro Evolution 121
Kaito Seo, Katsumi Hagino, and Norikazu Ichihashi

Rapid and Finely-Tuned Expression for Deployable Sensing Applications 141
Alexandra T. Patterson and Mark P. Styczynski

Adv Biochem Eng Biotechnol (2023) 186: 1–28
https://doi.org/10.1007/10_2023_231

Published online: 17 July 2023

Biomanufacturing by In Vitro Biotransformation (ivBT) Using Purified Cascade Multi-enzymes

Yanmei Qin, Qiangzi Li, Lin Fan, Xiao Ning, Xinlei Wei, and Chun You

Contents

1 Introduction 3
2 Design Principles and Techniques for ivBT 4
  2.1 Design Principles 5
  2.2 Design Techniques 7
3 Enzyme Mining and Engineering 9
  3.1 Enzyme Mining 9
  3.2 Enzyme Engineering 12
4 Construction and Optimization of ivSEBs 14
  4.1 Construction of ivSEBs 14
  4.2 Optimization of Reaction Conditions 16

Y. Qin and X. Ning
University of Chinese Academy of Sciences, Beijing, China

In Vitro Synthetic Biology Center, Tianjin Institute of Industrial Biotechnology, Chinese Academy of Sciences, Tianjin, China

Q. Li
In Vitro Synthetic Biology Center, Tianjin Institute of Industrial Biotechnology, Chinese Academy of Sciences, Tianjin, China

L. Fan
University of Chinese Academy of Sciences, Beijing, China

In Vitro Synthetic Biology Center, Tianjin Institute of Industrial Biotechnology, Chinese Academy of Sciences, Tianjin, China

University of Chinese Academy of Sciences Sino-Danish College, Beijing, China

X. Wei (✉) and C. You (✉)
In Vitro Synthetic Biology Center, Tianjin Institute of Industrial Biotechnology, Chinese Academy of Sciences, Tianjin, China

National Technology Innovation Center of Synthetic Biology, Tianjin, China
e-mail: wei_xl@tib.cas.cn; you_c@tib.cas.cn

5 Scale-up of ivBT 17
6 Perspectives 18
References 19

Abstract In vitro biotransformation (ivBT) refers to the use of an artificial biological reaction system that employs purified enzymes for the one-pot conversion of low-cost materials into biocommodities such as ethanol, organic acids, and amino acids. Unshackled from cell growth and metabolism, ivBT exhibits distinct advantages compared with metabolic engineering, including but not limited to high engineering flexibility, ease of operation, fast reaction rate, high product yields, and good scalability. These characteristics position ivBT as a promising next-generation biomanufacturing platform. Nevertheless, challenges persist in the enhancement of bulk enzyme preparation methods, the acquisition of enzymes with superior catalytic properties, and the development of sophisticated approaches for pathway design and system optimization. In alignment with the workflow of ivBT development, this chapter presents a systematic introduction to pathway design, enzyme mining and engineering, system construction, and system optimization. The chapter also proffers perspectives on ivBT development.

Graphical Abstract

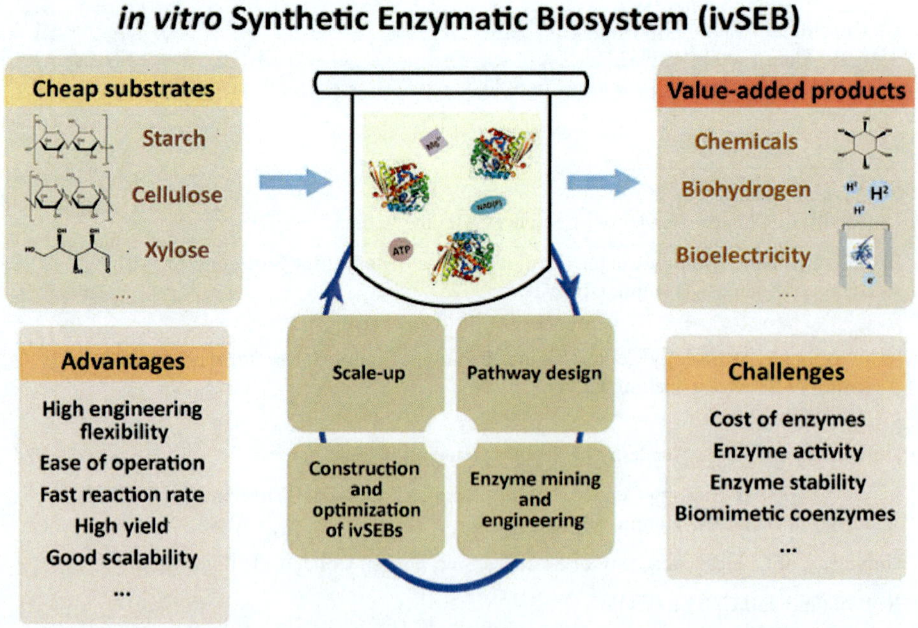

Keywords Biomanufacturing, Multi-enzymatic catalysis, Pathway design, Purified enzymes, Synthetic biology

1 Introduction

Amidst the depletion of natural resources, particularly petroleum, the use of renewable resources for biomanufacturing has garnered widespread interest. A number of microorganisms, including *Escherichia coli*, *Bacillus subtilis*, and *Saccharomyces cerevisiae*, have been genetically modified to produce an array of products ranging from primary to secondary metabolites [1]. Although microbial fermentation has achieved great success in biomanufacturing, it is still facing challenges including unpredictable engineering results (due to the intricate regulatory mechanisms of cells), low product yield, low titer of cell-toxic products, and slow mass transfer rate [2]. As an alternative biomanufacturing approach, cell-free synthesis (CFS) in bioprocessing traces its origins back to the 1890s when Eduard Buchner discovered that yeast extract could convert glucose to ethanol [3]. However, the milestone of in vitro biotransformation (ivBT), which is another biomanufacturing approach using in vitro synthetic enzymatic biosystems (ivSEBs), did not appear until the 1960s–1970s when a cocktail of amylase and amyloglucosidase was used to produce glucose from starch, and glucose isomerase could also be involved to isomerize glucose into fructose for the production of high-fructose corn syrup [4]. These pioneering works opened the door for ivBT. In recent times, driven by the advances in biochemistry, molecular biology, and enzyme engineering technologies, numerous enzymes with enhanced or novel features have been identified and characterized, making ivBT a promising method for the production of chemicals [5, 6].

We would like to clarify in this chapter that ivBT is distinct from CFS in terms of the aimed products and the catalysts, despite both being biomanufacturing approaches that do not use cells. CFS specifically refers to the utilization of cell lysates or purified recombinant elements for the production of nucleic acids or proteins [7] and is predominantly employed in the synthesis of RNAs [8], urgent vaccines [9], antibody–drug conjugates [10], and so on. In contrast, ivBT mediated by ivSEBs refers to the employment of multiple natural or artificial enzymes in purified forms (and sometimes with natural or biomimetic coenzymes) in a single vessel to produce biocommodities from renewable substrates such as starch, cellulose, and carbon dioxide (CO_2) (Fig. 1) [11, 12]. In many cases, ivBT also involves the use of electrodes [13–15] and/or organelles [16]. Unlike biomanufacturing by metabolic engineering that needs to compromise on cell growth and reproduction, ivBT takes biomanufacturing as its sole purpose. Therefore, ivBT can use the principle of simplifying complexity to reconstruct artificial metabolic pathways to guide the assembly of ivSEBs in a demand-driven manner, which makes it resemble a machine-based biomanufacturing approach. The advantages of ivBT for biomanufacturing, such as high product yield, fast reaction rate, and a wide range of reaction conditions, have been extensively discussed in many review papers [2, 6, 11, 17]. Additionally, ivBT is able to alleviate the limitations of microbial fermentation, realizing some important biotransformation processes that cannot be implemented by microorganisms such as the conversion of cellulose to starch [11]. Thus, the use of ivBT for the production of various important products is of

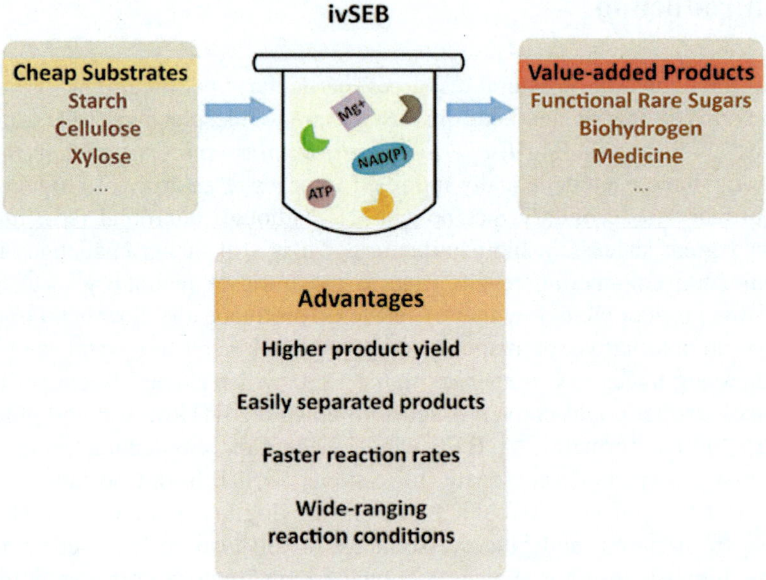

Fig. 1 Scheme of in vitro biotransformation (ivBT) mediated by an in vitro synthetic enzymatic biosystem (ivSEB)

immense and far-reaching importance. Nowadays, products that have been synthesized via ivBT include but not limited to functional rare sugars [18–20], sugar alcohols [21–23], organic acids [24–26], biomacromolecules [27–29], amino acid [15], glucosamine [30], bioelectricity [13, 14, 31, 32], and biohydrogen [33–36].

In the subsequent sections, we will concentrate on elucidating the fundamental design principles and techniques for ivBT. Moreover, we will underscore the significance of enzyme mining, enzyme engineering, and system optimization as key approaches to construct ivSEBs with high biomanufacturing efficiencies.

2 Design Principles and Techniques for ivBT

The development of ivBT following a cyclic workflow consists of five stages: (1) pathway design, (2) enzyme selection, (3) enzyme engineering, (4) enzyme production, and (5) process engineering [11]. Among these stages, pathway design is considered the crucial step for ivBT development. The general design principles and techniques, concluded from previously reported successful examples, are discussed in this section.

2.1 Design Principles

Given that the same biocatalytic reaction can often be accomplished by multiple different enzyme cascades, it is of great importance to carefully design the pathways for ivBT. Various factors must be considered during the design process, such as substrate cost, thermodynamics, coenzymes requirements (e.g., ATP, NAD(P)H), and coenzyme balance [11]. The ensuing discussion will provide illustrative examples.

2.1.1 Cheap Substrate

As introduced in Sect. 1, ivBT aims at the production of biocommodities rather than high-value products such as drug precursors and antibiotics. Therefore, the use of low-cost and renewable substrates for ivBT is of utmost importance to ensure the economic feasibility of biomanufacturing. Examples of such substrates include glucose, xylose, whey waste, CO_2, oligosaccharides, and polysaccharides. Among these, oligosaccharides (e.g., sucrose and cellobiose) and polysaccharides (e.g., cellulose and starch) are ideal substrates for ivBT. Energy in the glycosidic bonds of these saccharides can be conserved by phosphorylases to generate sugar phosphate which can be utilized downstream, without the need for external energy input such as ATP [22, 27, 33, 37, 38]. Additionally, the readily available and inexpensive CO_2 is also an attractive substrate for ivBT [24, 39]. The utilization of CO_2 for ivBT will be in line with the global carbon reduction goals and promotes the implementation of low-carbon initiatives.

2.1.2 Thermodynamic Analysis

When designing the pathways for ivBT, thermodynamics is another key issue to consider [11]. One approach to perform thermodynamic analysis is to calculate the change in Gibbs free energy ($\Delta_r G'^{\circ}$) of the reaction on the eQuilibrator website (http://equilibrator.weizmann.ac.il) [40, 41]. A negative overall $\Delta_r G'^{\circ}$ value is a crucial criterion for an ivBT pathway that is worth of experimental implementation. Furthermore, the last catalytic step of the pathway should preferably be irreversible to drive the overall reaction toward the direction of product generation, ensuring high theoretical product yields. Examples of this type of ivBT include those for the production of inositol [21], allulose [19], glucosamine [30], fructose [42], tagatose [43, 44], mannose [45], and mannitol [23]. The pathways of these examples share a common feature, in which the last step is an irreversible dephosphorylation reaction catalyzed by a phosphatase. Additionally, ivBT for the synthesis of gaseous products (such as hydrogen [34, 36]) or insoluble products (such as cellulose [29], amylose [27, 39], and PHB [28, 29]) often results in high product yields, because these products can be readily removed from the reaction systems in situ.

2.1.3 The Balancing and Regeneration of Coenzymes

While it is ideal to design coenzyme-free pathways for ivBT, sometimes the involvement of coenzymes is necessary. In such cases, the ivBT pathways should be coenzyme-balanced. Coenzyme-balanced ivBT pathways can be classified into two types. The first type of pathway generates and consumes coenzymes stoichiometrically by the enzymes which belong to the pathway, and do not result in the formation of by-products. For example, Sieber and his team designed two NAD(H)-balanced ivBT pathways in which NADH generated from glucose catalyzed by glucose dehydrogenase (GDH) was utilized stoichiometrically in the last step catalyzed by different alcohol dehydrogenases to produce ethanol and isobutanol, respectively [46]. Building upon this study, You's team designed an NAD(H)-balanced ivBT pathway for the production of lactate from glucose [25]. In another study, Bowie's team proposed an ATP-balanced ivBT pathway for the production of poly(3-hydroxybutyrate) (PHB) from glucose [28]. This same group of researchers also designed another ivBT pathway for the production of isobutanol from glucose [47]. The ivSEB assembled based on this pathway not only achieved the balance of NADPH but also maintained the ATP level through a molecular rheostat module that could regulate ATP concentrations in response to inorganic phosphate concentration in the system. The second type of ivBT pathway necessitates the use of external modules containing exogenous coenzyme donors and the corresponding oxidoreductases for coenzyme regeneration, which is accompanied with the generation of by-products. These external modules consist of substrates/enzymes such as formate/formate dehydrogenase [23, 48], glucose/GDH [49, 50], phosphite/phosphite dehydrogenase [51], and hydrogen/hydrogenase [52] for the regeneration of NAD(P)H from NAD(P)$^+$, and oxygen/NAD(P)H oxidases [28] for the regeneration of NAD (P)$^+$ from NAD(P)H. ATP regeneration can be achieved using high-energy phosphate-containing substrates such as acetyl phosphate, creatine phosphate, phosphoenolpyruvate, and polyphosphate. Among these, polyphosphate is most commonly used due to its affordability [53]. However, high concentrations of polyphosphate can chelate with magnesium ions in the reaction system, leading to reduced activity of magnesium-dependent enzymes [54]. This issue also arises with the accumulation of phosphorus-containing by-products including orthophosphate, pyrophosphate, and triphosphate resulting from a prolonged reaction of longer-chain polyphosphate by polyphosphate kinase for ATP regeneration [22, 55]. To overcome this issue, phosphate-free substrates such as starch, glucose, and pyruvate can be utilized for ATP regeneration via cascade enzymatic reactions. For instance, Kim and Swartz devised an enzymatic pathway that generates one ATP molecule from every pyruvate molecule [56]. This pathway, named the PANOx system, was integrated into a more complex enzymatic pathway designed by Zhang and coworkers to produce 4 ATPs from 1 glucose unit of starch [57]. To circumvent the use of NAD(H) as in the PANOx system, You's team designed an NAD(P)H-free ATP regeneration system that produces 3 ATPs from 1 glucose unit of starch [58]. However, the usage of these systems for ATP regeneration inevitably leads to the generation of by-products such as acetate. Additionally, hybrid catalytic systems incorporating electricity and/or light can also be applied to ivBT for coenzyme regeneration without the by-products. For instance, Zhu's group developed a hybrid CO_2 electroreduction system by coupling the reaction of FDH for converting CO_2 to formate with NADH regeneration catalyzed by an electrode modified with copper nanoparticles [59]. The same group also employed thylakoid membranes isolated from spinach for the light-driven co-regeneration of ATP and NADPH, and utilized this strategy for the production of PHB from acetate [16]. Jiang's team designed an artificial thylakoid by coating the inner wall of titania microcapsules with cadmium sulfide quantum dots, and coupled this NADH-regenerating artificial thylakoid with an NADH-consuming enzyme cascade for the reduction of CO_2 to methanol [60].

2.2 *Design Techniques*

In many cases, pathways for ivBT are designed according to natural metabolic pathways with necessary modifications [11]. Yet apart from using nature-derived biocatalytic pathways, the de novo design of artificial ivBT pathways can offer advantages such as fewer catalytic steps and/or improved catalytic efficiency. The following subsections will explore these two design techniques in detail.

2.2.1 Reconstruction of Natural Metabolic Pathways

Natural metabolic processes such as glycolysis and the pentose phosphate shunt have served as inspirations for the design of various ivBT pathways. For instance, utilizing the upper glycolytic pathway consisting of hexokinase, phosphoglucose isomerase, phosphofructokinase, aldolase, and triose phosphate isomerase, various ivBT pathways have been designed for the production of sugar phosphates such as ketose 1-phosphates [61] and xylulose 5-phosphate [62] from glucose. In many cases, the designed ivBT pathways overlap significantly with natural metabolic pathways, but novel non-natural reactions are also incorporated, serving as the key for implementing the entire pathway. An example of this is the ivBT pathway designed by Zhang and his team for the production of biohydrogen from xylo-oligosaccharides [35]. The upstream part of this pathway was an ATP-free artificial reaction cascade to produce xylulose 5-phosphate (Xu5P) from xylo-oligosaccharides based on the discovery of novel activities of cellodextrin phosphorylase, phosphopentomutase, and ribose 5-phosphate isomerase. Xu5P produced by this new cascade reaction could be converted to glucose 6-phosphate (G6P) through the natural pentose phosphate pathway (PPP) downstream. The PPP pathway could stoichiometrically regenerate 12 NADPH molecules from each G6P molecule, and the NADPH was subsequently used by a hydrogenase to produce hydrogen. With this design, the whole ivBT pathway demonstrated a high theoretical yield of 10 H_2 per xylose unit of xylo-oligosaccharides.

2.2.2 In Silico Design of Artificial ivBT Pathways

In recent years, computational approaches have emerged as a significant tool in the design of artificial ivBT pathways. The in silico pathway design workflow typically entails five steps: (1) the creation of a database that allows for the selection of enzymatic reactions and metabolites to constitute the pathway; (2) the establishment of the curated reactions and metabolite database for pathway search, which can be in the form of either a graph or a stoichiometric matrix; (3) network pruning, which can be achieved through various strategies such as the exclusion of coenzymes due to their ubiquitous participation in metabolic networks, as well as stoichiometry-based approaches like flux balance analysis (FBA); (4) the selection of search algorithms; and (5) pathway ranking, which can be based on the criteria such as number of reaction steps, availability of natural enzymes, thermodynamic feasibility, and product yield [63].

An example of computational pathway design is the drafting of an artificial chemical–biochemical hybrid pathway for the fixation of CO_2 into starch [39]. In this study, Ma and colleagues opted for formic acid and methanol as starting substrate candidates for the pathway [39], because these compounds can be produced from CO_2 through chemical approaches in high efficiency [64, 65]. Thereafter, possible ivBT pathways from either formic acid or methanol to starch were proposed by the COBRApy toolbox in Python [66], using a combination of combinatorial algorithm and FBA algorithm [67]. The reaction set consisted of over 6,500 reactions extracted from the MetaCyc database [68] and the ATLAS database [69], along with additional reactions utilizing formic acid or methanol. Two linear, short, and carbon-conserving ivBT pathways were selected for the synthesis of starch from formic acid or methanol, respectively. These two pathways were subsequently broken down into multiple reaction modules for experimental validation and combination, ultimately resulting in an 11-step ivBT pathway for the efficient conversion of methanol to starch [39].

In recent times, the advance of artificial intelligence (AI) technology and the proliferation of biological big data have facilitated the emergence of data-driven methods such as deep learning and reinforcement learning, which have further enriched the repertoire of pathway design techniques [70]. An example of the successful application of such techniques is the development of RetroPath RL (https://github.com/brsynth/RetroPathRL) by Faulon and colleagues, a retrosynthetic program that employs the Monte Carlo Tree Search reinforcement learning algorithm for designing biosynthetic pathways aimed at a specific target product [71]. The efficacy of RetroPath RL was demonstrated by its successful prediction of 127 out of 152 experimentally validated pathways, thus revealing its potential to facilitate the design of experimentally viable pathways.

3 Enzyme Mining and Engineering

Following pathway design, the subsequent step toward the development of ivBT is the selection of suitable enzymes for implementing the designed pathway [11]. To enhance the adaptability of enzymes in the ivSEBs, those enzymes with high catalytic activity, excellent substrate specificity, low product inhibition, and desirable stability are preferred [11, 72]. In instances where existing enzymes are unsuitable for the designed ivBT approach, novel enzymes with superior desired properties can be obtained through enzyme mining and enzyme engineering (Fig. 2).

3.1 Enzyme Mining

In the past few decades, significant progress in DNA technology, protein technology, and bioinformatics has resulted in an enormous amount of data and facilitated the development of diverse enzyme mining strategies. A prevalent method for

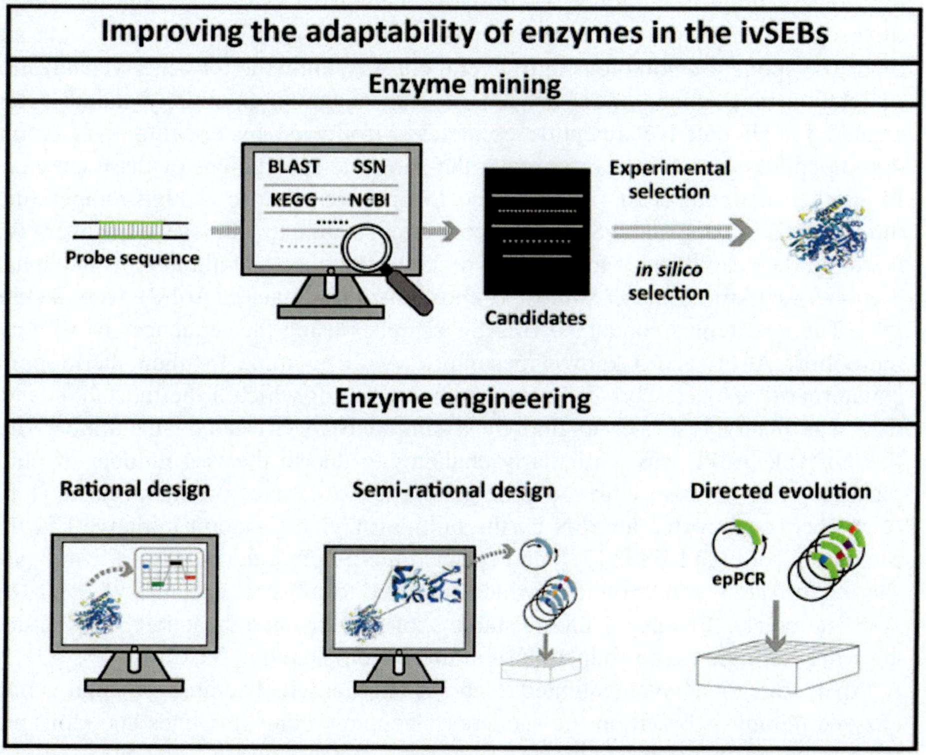

Fig. 2 Methods for improving enzyme adaptability in the ivSEBs, including enzyme mining and enzyme engineering

enzyme discovery is based on the searching for gene or protein sequences with known functions in databases such as NCBI, Pfam, and InterPro. This approach involves using a template sequence, known as a probe, to search for homologous sequences in databases. For example, Balskus, Turnbaugh, and colleagues found a new enzyme with tyrosine/levodopa decarboxylase activity in *Enterococcus faecalis* using the sequence of tyrosine decarboxylase from *Lactobacillus brevis* as a probe through a BLAST (Basic Local Alignment Search Tool) search within the Human Microbiome Project reference genomes [73]. Similarly, Zhu's group found a putative aldehyde reductase (AR) from *Oceanospirillum* sp. through the BLAST search with the sequence of AR from human liver as a probe [74]. This enzyme was subsequently characterized and found to catalyze the chemoselective reduction of aldehydes in the presence of ketones, highlighting its potential for biosynthesis. Enzyme mining using a probe sequence is a simple method, but may result in enzymes with unsatisfactory properties such as low catalytic specificity and poor thermostability.

Hence, in addition to the utilization of sequence information, the use of structural information for analysis can improve the likelihood of finding new enzymes, particularly for revealing the enzyme substrate specificity [75]. For example, a dipeptide epimerase from *Bacteroides thetaiotaomicron* was identified based on the construction of a sequence similarity network (SSN) and high-throughput molecular docking [76]. The researchers conducted an SSN analysis to classify dipeptide epimerase homologs into several clusters, allowing for better visualization of relationships among these sequences. Subsequently, homology models were created for all putative dipeptide epimerases, followed by docking with around 400 dipeptides as substrates to predict the substrate specificities of these enzymes. In another example, You's team aimed to produce allulose at high temperatures through ivBT using an ivSEB that converted starch to allulose. However, two thermophilic candidates for the two key enzymes, D-allulose 6-phosphate 3-epimerase (A6PE) and D-allulose 6-phosphate phosphatase (A6PP), were lacking [19]. The researchers conducted BLAST searches using the sequences of a known mesophilic A6PE and a known mesophilic A6PP to mine for their thermophilic counterparts, respectively. Compared with A6PE, for which a thermostable candidate was relatively easier to find by a simple BLAST search, the mining of a thermostable A6PP was particularly challenging due to the vast number of phosphatases with diverse substrate promiscuities. To ensure mining efficiency, the researchers constructed an SSN by the online Enzyme Function Initiative-Enzyme Similarity Tool (EFI-EST) [77], and selected thermophile-derived A6PP candidates sharing the same "cap" structure (which is related to substrate specificity [78]) as the A6PP template. Finally, a thermostable A6PP with enough activity was found, allowing for high-yield production of allulose from starch at 55°C.

Apart from the abovementioned methods, another widely employed approach for enzyme mining is based on the sequenced genomic context. Genes encoding proteins involved in the production of a specific metabolite are frequently found in proximity to one another in biosynthetic gene clusters (BGCs), thereby allowing for the prediction of the function of an uncharacterized enzyme based on the genes located in the same locus [75]. Several computational tools have been developed for the identification of gene clusters, such as EFI-GNT, PULDB, ClusterFinder, and STRING [79]. For instance, Jiang's team utilized the webtool STRING to identify potential interacting partners of two genes known to participate in flavone metabolism in *Flavonifractor plautii*, ultimately discovering a flavone reductase from this microorganism [80]. Nevertheless, the genomic context is influenced by the factors such as gene gain/loss/rearrangement during evolution [81], which can complicate the use of this method for enzyme mining.

The giant gap between the numbers of sequences and the experimental results on enzyme functions highlights the significance of computational approaches for enzyme annotation and mining. In the abovementioned example of mining a new dipeptide epimerase, computational algorithms for homology modeling and docking were applied [76]. Liao, Siegel, and colleagues developed a combinatorial approach based on bioinformatics and molecular modeling to mine for a ketoisovalerate decarboxylase (KIVD) that could be readily expressed in *E. coli* and could specifically use long-chain ketoacids as substrates [82]. These researchers first identified more than 2,000 KIVD candidates through sequence alignment with a known KIVD, followed by the application of a sequence identity cutoff of 90% and the removal of eukaryotic enzymes to reduce the number of candidates. Homology models of these candidates were then generated using Rosetta Comparative Modeling [83], followed by a comparison with the native KIVD crystal structure using the TMalign algorithm [84] to further narrow down the number of KIVD candidates. For the 239 remaining KIVD candidates, a C8 substrate was docked into their predicted active sites, and interface energy calculations were performed by Rosetta Design to predict their activities for the C8 substrate. This method led to the finding of a new KIVD with a more than 100-fold improvement in substrate specificity (C8 versus C5) compared with the KIVD template. However, when applied in vivo as a key enzyme for the production of long-chain alcohols from glucose, this newly found KIVD resulted in a significant drop in product titer compared with the KIVD template enzyme, possibly due to the cell-toxic nature of longer-chain alcohols. The productivity may be enhanced by applying the same pathway and enzymes to ivBT.

The development of AI technology has shown promise in the field of protein function prediction. Machine learning models, in particular, have demonstrated good predictive performance [85]. For example, Atalay, Doğan, and colleagues developed ECPred, which used a machine learning algorithm to predict enzymatic functions based on the Enzyme Commission (EC) nomenclature [86]. Ferrari's team also employed multi-label machine learning to predict enzymatic functions at the level of chemical mechanism [87], providing predictions with greater details such as reaction mechanism and coenzyme requirement compared to the EC nomenclature-based approach.

3.2 *Enzyme Engineering*

Despite their ability to operate under mild reaction conditions and exhibit relatively high catalytic selectivity, enzymes may possess certain limitations such as inadequate catalytic activity, instability under operating conditions, poor stereoselectivity, poor regioselectivity, unsatisfactory substrate scope, and susceptibility to product inhibition [88]. To address these issues, enzyme engineering approaches, including rational design, semi-rational design, and directed evolution, are commonly employed. The selection of an appropriate engineering approach is dependent upon the desired properties of the target enzyme, the availability of relevant structural and functional information, and the availability of a suitable high-throughput screening or selection method [89].

3.2.1 Rational Design

Rational design is an enzyme engineering approach that relies on the knowledge of protein structures and mechanisms [90]. Unlike directed evolution (as discussed in Sect. 3.2.3), rational design involves modifying specific amino acid residues of an enzyme using site-directed mutagenesis to generate a limited number of mutants for subsequent validation. In recent years, a variety of molecular modeling techniques (such as SWISS-MODEL [91], I-TASSER [92], and AlphaFold [93]) have been developed to predict protein structures based on amino acid sequences. Furthermore, significant progress has been made in the field of simulation and molecular docking of small molecules to proteins [94, 95]. These approaches provide valuable support for rational design even in the absence of crystal structures, resulting in the successful improvement of specific activity [96, 97], thermostability [98, 99], substrate specificity [100, 101], and other properties of many enzymes.

An example of the application of rationally deigned enzymes to enhance the performance of ivSEBs is the development of an enzymatic biobattery by Zhang's group [102]. In this study, the objective was to modify the coenzyme specificity of a thermophilic 6-phosphogluconate dehydrogenase (6PGDH) from $NADP^+$ to NAD^+. By aligning the protein sequences of $NADP^+$- and NAD^+-preferred 6PGDHs and conducting in silico docking of the substrate with the coenzyme, the crucial amino acid residues involved in the interaction with the phosphate group of $NADP^+$ were identified. Subsequently, four mutants were constructed, among which the most effective mutant exhibited a reversal of the coenzyme specificity from $NADP^+$ to NAD^+ by around 64,000 folds. The biobattery incorporating this 6PGDH mutant displayed a maximum power density and current density approximately 25% higher than the biobattery with the wild-type enzyme.

3.2.2 Semi-rational Design

Semi-rational design is an enzyme engineering approach that combines rational design (Sect. 3.2.1) and directed evolution (Sect. 3.2.3) [103]. This approach involves selecting several residues for mutation based on structural or functional information, creating "small and smart" mutant libraries for screening or selection [104]. Semi-rational design is particularly useful for modifying enzyme activities because mutations at active sites frequently exhibit synergistic effects [104, 105]. For example, Zhang and colleagues altered the coenzyme preference of glucose 6-phosphate dehydrogenase (G6PDH) from $NADP^+$ to NAD^+ through a semi-rational approach [36]. They began by constructing a homology model of the G6PDH to be engineered, followed by saturation mutagenesis of four key amino acid residues for binding the phosphate group of $NADP^+$. Three mutant libraries were constructed using the conventional NNK codon degeneracy method, and were screened for the mutants with enhanced NAD^+-dependent activities. The best mutant was obtained through a combination of positive mutants and applied to an NAD^+-dependent ivSEB for the production of biohydrogen from starch. Other codon degeneracies, such as the NDT degeneracy, can be utilized to reduce the screening workload. The NDT degeneracy method employs only 12 codons to code a structurally balanced set of 12 amino acids, significantly reducing the number of mutants to be screened [106, 107]. Semi-rational design can also be implemented using computational approaches. Hellinga and colleagues developed a computational algorithm for designing enzyme active sites, successfully transforming a ribose-binding protein into a triose phosphate isomerase [108]. Their study demonstrated the feasibility of creating novel enzyme activities in non-catalytic protein scaffolds.

3.2.3 Directed Evolution

The process of directed evolution emulates the principles of natural evolution. Initially, the gene encoding the enzyme of interest is subjected to random mutagenesis by error-prone polymerase chain reaction (epPCR), resulting in a library of enzyme mutants. This mutant library is then introduced into a microbial host to express the enzymes. Subsequently, the screening or selection process is conducted to isolate enzyme mutants with improved properties. The screening approach utilizes analytical tools such as UV/Vis spectrophotometry, fluorescence, high-performance liquid chromatography (HPLC), and gas chromatography (GC) to identify enzyme mutants with the desired properties. In contrast, in the selection process, the host organism expressing an enzyme mutant with desired catalytic properties gains growth and survival advantages. Finally, the gene of the best mutant is extracted and utilized as a template in the next round of directed evolution [88].

Directed evolution is a highly effective technique for enhancing the stability, activity, and substrate selectivity of enzymes, and is often preferred over rational design for modifying enantio-, diastereo-, and/or regioselectivity of enzymes

[109]. Moreover, directed evolution is well suited for engineering enzymes that lack structural information. However, identifying the desired mutants from a library with a vast number of mutants can pose a significant challenge, making the development of high-throughput screening or selection methods crucial. Zhang and coworkers proposed a double-layer high-throughput screening method for identifying mutants of a thermostable 6-phosphogluconate dehydrogenase (6PGDH) with altered coenzyme selectivity from $NADP^+$ to NAD^+ [110]. This method involves heat treatment of colonies of a 6PGDH mutant library on agar plates to disrupt the cell membrane and minimize background noise due to intracellular dehydrogenases and inherent NAD(P)H, followed by the addition of melted agarose solution containing tetranitroblue tetrazolium (TNBT), phenazine methosulfate (PMS), NAD^+, and 6-phosphogluconate to form a second layer. The reduced NADH produced by more active 6PGDH mutants reacts with TNBT in the presence of PMS, yielding a black color that is easily identified by naked eye. Similar screening methods have been applied to the engineering of other thermophilic redox enzymes [111, 112] and thermophilic enzymes that can produce NAD(P)H when coupled with a cascade enzyme [113]. In silico screening, aided by machine learning approaches, can also enhance the efficiency of directed evolution [114, 115]. For example, through traditional directed evolution in combination with a machine learning model, Gjalt and colleagues improved the volumetric productivity of a halohydrin dehalogenase by around 4,000-fold [116].

4 Construction and Optimization of ivSEBs

After selecting the enzymes suitable for the designed ivBT pathway, the subsequent step is to prepare these enzymes and their experimental integration into an ivSEB for biosynthesis. Typically, enzymes are expressed in *E. coli*, *B. subtilis*, and yeast, and subsequently purified through cost-effective methods such as heat precipitation and adsorption [11]. Then, a proof-of-concept trial is carried out, followed by optimizing the reaction conditions for satisfied reaction rate, product yield, and product titer [72]. In this section, we will provide an overview of the methods used for the construction and optimization of ivSEBs.

4.1 Construction of ivSEBs

When constructing an ivSEB, enzymes can be applied in the form of free enzyme cocktails, multi-enzyme complexes, or immobilized enzymes (Fig. 3). The advantages and disadvantages of each form of enzymes are discussed in the following subsections.

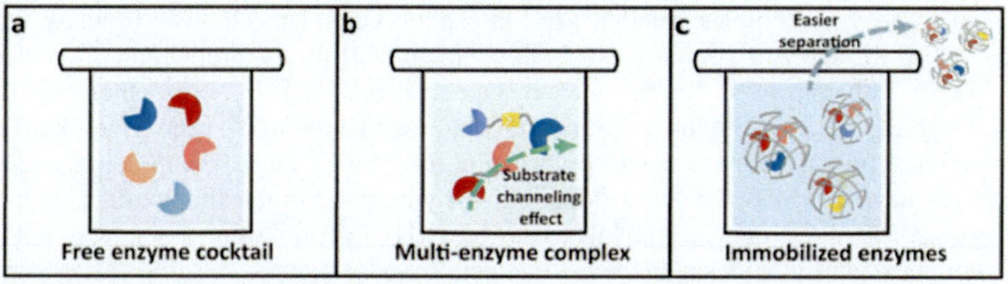

Fig. 3 Three types of ivSEBs that use different forms of enzymes. (**a**) ivSEB that uses a free enzyme cocktail, (**b**) ivSEB that uses a multi-enzyme complex, (**c**) ivSEB that uses immobilized enzymes

4.1.1 Free Enzyme Cocktails

During the early stages of development, free enzyme cocktails are frequently employed for ivSEBs [21, 28]. In this method, enzymes are usually expressed and purified individually, and mixed in one pot with the supplementation of buffer, substrates, and other ingredients such as metal ions and coenzymes to implement the reactions. The prepared enzymes can be frozen in solution or in powder form for prolonged storage, allowing for their flexible utilization in the construction of diverse ivSEBs. However, the use of free enzyme cocktails is not without challenges, including enzyme instability, low intermediate transfer efficiency, and difficulties with product separation and enzyme reuse [117, 118]. These issues may be mitigated through the use of multi-enzyme complexes and/or immobilized enzymes.

4.1.2 Multi-enzyme Complexes

In natural system, cascade enzymes are often spatially colocalized for the efficient transportation of reaction intermediates [119]. This has inspired the development of various methods for constructing artificial multi-enzyme complexes to improve the biocatalytic efficiency. For example, Zheng, Sun, and coworkers have constructed an ivSEB for the conversion of methanol and ribose 5-phosphate into fructose 6-phosphate (F6P) using a three-enzyme fusion, which demonstrated a faster F6P production rate than the ivSEB using a free enzyme cocktail [120]. Similarly, based on the covalent SpyTag–SpyCatcher interaction [121] and the noncovalent cohesin–dockerin interaction [122, 123], You, Zhang, and coworkers constructed a four-enzyme ivSEB for the conversion of starch to inositol [124]. This enzyme complex exhibited higher initial reaction rates than the free enzyme cocktail, and was advantageous for its straightforward preparation by mixing crude cell extracts.

For the construction of more complicated ivSEBs, the enzyme catalyzing the rate-limiting step is specifically selected along with its downstream enzyme(s) to form a complex. This enzyme complex can then be combined with other enzymes in free cocktail form to construct the designed ivSEB. For instance, You's group assembled a two-enzyme complex for the faster phosphorylation of cellulosic biomass, and applied this enzyme complex to a five-enzyme ivSEB for the production of bioelectricity with improved current and power densities [125]. In some other ivSEBs, enzymes involved in the consumption and regeneration of coenzymes can be selected for assembly to enhance coenzyme cycling. Zhang's group constructed a five-enzyme ivSEB for the production of hydrogen from maltodextrin [126], in which the three enzymes involved in NAD(H) cycling were assembled into a complex via the cohesin–dockerin interaction and exhibited 4.5-fold faster initial reaction rate than the free enzyme cocktail.

4.1.3 Immobilized Enzymes

In contrast to free enzyme cocktails, immobilized enzymes have been shown to offer improved stability and reusability [127]. Commonly used immobilization approaches can be categorized into three types: adsorption on a support material (the carrier), encapsulation in a compartment, and chemical cross-linking [127]. An example of physical adsorption is the use of porous polydopamine microspheres for the co-immobilization of all four enzymes in an ivSEB for the conversion of starch into inositol [128]. This co-immobilization method resulted in a reaction rate comparable to that of the free enzyme cocktail, while showing much higher thermostability and recovery stability. A more recent study following the same inositol-producing ivBT pathway used biomimetic mineralized microcapsules made of silicate and polyethyleneimine for the co-encapsulation of the four enzymes, resulting in a half-life of enzymes in this microcapsule 5.9 times as that of the free enzyme cocktail [118]. Moreover, by utilizing co-encapsulated enzymes, this designed ivSEB achieved a high inositol titer of 210 g/L, demonstrating its potential for industrial applications. Chemical cross-linking has also been applied to immobilize enzymes in ivSEBs, such as the use of glutaraldehyde as the cross-linking reagent for the co-immobilization of three enzymes to magnetic nanoparticles for one-pot hydrolysis of starch into glucose [129].This tri-enzyme ivSEB remained completely active after eight reaction cycles with a cumulative time of 12 h at 70°C, possibly due to the covalent bonds formed between these enzymes and the magnetic nanoparticles that helped to restrict molecular flexibility.

4.2 Optimization of Reaction Conditions

Besides the utilization of enzymes with high catalytic activities and high stability, it is also crucial to identify appropriate reaction conditions for the adaptation of enzymes in the ivSEBs [72]. Because enzymes in an ivSEB are often derived from different organisms, they may exhibit varying optimal pH values, optimal temperatures, optimal ionic strengths, cofactor preferences, requirements of metal ions as activators, substrate specificities, etc. [11]. Therefore, a compromise must be reached

[11]. For instance, Lou and coworkers developed a seven-enzyme ivSEB for the one-pot conversion of sucrose into glucaric acid (GA) [130] and optimized the reaction conditions by testing various factors, such as temperatures, pH, buffer type and concentration, NAD^+ concentration, sucrose concentration, and enzyme loading amounts, ultimately achieving an improvement of product yield from around 20 to 75% of the theoretical value. Similar strategies have been applied for the optimization of other ivSEBs designed for the high-yield production of lactate [25], laminaribiose [18], glucosamine [30], mannitol [23], and fructose 1,6-diphosphate [131]. For complicated ivSEBs comprised of ten or more enzymes, however, single-factor optimization will be laborious. In such cases, a strategy involving the simultaneous reduction of enzyme concentrations to 1/5 or 1/25 of their original values could quickly help to identify rate-limiting enzymes in the system, as demonstrated by Bowie and his team [28].

With the rapid development of bioinformatics and computational technologies, in silico analysis using kinetic models is becoming a promising strategy for the optimization of ivSEBs. For example, Zhang's team built a kinetic model for their three-enzyme ivSEB to predict the best enzyme loading ratio and the optimal phosphate concentration [132], while You and colleagues developed a kinetic model for optimizing the reaction conditions of another three-enzyme ivSEB designed for the production of rare disaccharides from starch [20]. Despite these successful examples, it is still a great challenge at present to construct an accurate computational model for predicting the performance of a more complex ivSEB.

In addition to the aforementioned methods, other strategies for optimizing reaction conditions include (1) the implementation of one-pot, multi-step reactions to avoid accumulation of toxic reaction intermediates [39] or to accommodate varying temperature preferences of enzymes in the same ivSEB [22], (2) fed-batch addition of substrates to prevent enzyme inhibition, which is often necessary in ivSEBs that use pyrophosphate or polyphosphate for the regeneration of ATP [37, 131], and (3) in situ removal of products to prevent enzyme inhibition [133].

5 Scale-up of ivBT

Unlike microbial fermentation, which encounters obstacles in scaling up from laboratory to industrial conditions due to heterogeneous fermentation conditions [134], ivBT allows for easier scale-up because ivSEBs exhibit similar behavior at both laboratory and larger scales [17, 135]. Therefore, ivBT presents a competitive alternative for biomanufacturing. An example of successful industrial biomanufacturing by ivBT was the production of inositol from starch operated in 20,000-L reactors in 2017 by Tianjin Institute of Industrial Biotechnology (TIB), Chinese Academy of Sciences (CAS) [21]. Several features facilitated the successful industrialization of this ivBT approach, including the use of low-cost starch as the substrate, the design of a simple, thermodynamically favorable, coenzyme-free, and by-product-free pathway composed of only five enzymes, the selection of hyperthermophilic enzymes that could be purified by the low-cost heat precipitation, and the capacity of thermophilic enzymes to operate at high temperatures to avoid microbial contamination. The production cost of inositol by ivBT was roughly one-third that of the traditional phytate-based manufacturing method [21]. In addition to inositol, TIB has also carried out industrial production layout for rare sugars (such as tagatose) [136] and artificial starch [137].

The implementation of ivBT at large scales necessitates the development of approaches for economically feasible bulk enzyme purification, reaction process optimization, target product separation, and other pertinent factors [12]. Decreasing the cost of enzymes is crucial for the industrialization of ivBT [1]. There has been a wide concern that enzyme production is very expensive, given that it costs about $1,000 to obtain 1 mg of purified enzyme in laboratories using conventional methods including the shake-flask culture of *E. coli*, isopropyl-β-D-thiogalactopyranoside (IPTG) induction in Luria-Bertani (LB) medium, and nickel-affinity column chromatography [1]. In fact, however, bulk production of enzymes using high-cell-density fermentation of secretory hosts can be less costly, being as low as about $10–30/kg purified enzymes [1]. In recent years, Zhang and colleagues in TIB have been working diligently to develop low-cost enzyme immobilization technologies for decreasing the cost of enzymes, thereby to further decrease the industrial manufacturing cost of inositol. The cost of enzymes for producing 1 kg of inositol is now as low as $1.09 [118]. Meanwhile, this research team has developed a method for facile protein expression at high levels in *B. subtilis* by the use of an T7-promoter-based protein expression system [138]. This expression system featured its easy genetic operation, stable expression of target proteins at high levels, wide applicability, and its suitability for the bulk production of enzymes [138].

6 Perspectives

ivBT represents a highly promising platform for cell-free production, given its numerous advantages including high engineering flexibility, ease of operation, fast reaction rates, high product yields, and good scalability. To date, a diverse array of ivBT approaches have been developed for the production of biochemicals and bioenergy at near-theoretical yields [13, 14, 18, 21–23, 33, 35, 131, 139], with notable successes including the industrialized ivBT approach for inositol production from starch [21]. However, significant obstacles still impede the advancement of ivBT, including the cost of bulk enzyme preparation, the enhancement of enzyme activity and/or stability, standardization of enzymes and reaction modules, utilization of cheaper and more stable biomimetic coenzymes, and the scaling-up of processes [12]. Furthermore, one of the most important future trends regarding the development of ivBT may focus on the application of AI and machine learning approaches for the de novo design of enzymes [70, 114, 115, 140, 141] and reaction pathways [70, 71], the identification and annotation of new enzymes [85], and even the prediction of protein expression levels [142]. Through the integration of knowledge across biological, chemical, computational, mathematical, and engineering disciplines, we may be able to overcome these challenges and unleash the full potential of ivBT as a powerful biomanufacturing platform for the industrial production.

Acknowledgments We are grateful to Professor Yi-Heng P. Job Zhang, the director of in vitro Synthetic Biology Center of TIB, CAS, for providing scientific suggestions.

References

1. Zhang Y-HP, Sun J, Ma Y (2017) Biomanufacturing: history and perspective. J Ind Microbiol Biotechnol 44(4–5):773–784. https://doi.org/10.1007/s10295-016-1863-2
2. Claassens NJ, Burgener S, Vogeli B, Erb TJ, Bar-Even A (2019) A critical comparison of cellular and cell-free bioproduction systems. Curr Opin Biotechnol 60:221–229. https://doi.org/10.1016/j.copbio.2019.05.003
3. Buchner E (1897) Alkoholische Gärung ohne Hefezellen. Ber Dtsch Chem Ges 30:117–124
4. Fernandes P (2010) Enzymes in food processing: a condensed overview on strategies for better biocatalysts. Enzyme Res 2010:862537. https://doi.org/10.4061/2010/862537
5. Bornscheuer UT, Huisman GW, Kazlauskas RJ, Lutz S, Moore JC, Robins K (2012) Engineering the third wave of biocatalysis. Nature 485(7397):185–194. https://doi.org/10.1038/nature11117
6. Fessner WD (2015) Systems biocatalysis: development and engineering of cell-free "artificial metabolisms" for preparative multi-enzymatic synthesis. New Biotechnol 32(6):658–664. https://doi.org/10.1016/j.nbt.2014.11.007
7. Melinek B, Colant N, Stamatis C, Lennon C, Farid S, Polizzi K et al (2020) Toward a roadmap for cell-free synthesis in bioprocessing. BioProcess Int. https://doi.org/10.1016/B978-0-08-100623-8.00044-X
8. Pe'ery T, Mathews MB (1997) Synthesis and purification of single-stranded RNA for use in experiments with PKR and in cell-free translation systems. Methods 11(4):371–381. https://doi.org/10.1006/meth.1996.0435
9. Pardee K, Slomovic S, Nguyen PQ, Lee JW, Donghia N, Burrill D et al (2016) Portable, on-demand biomolecular manufacturing. Cell 167(1):248–259.e12. https://doi.org/10.1016/j.cell.2016.09.013
10. Stamatis C, Farid SS (2021) Process economics evaluation of cell-free synthesis for the commercial manufacture of antibody drug conjugates. Biotechnol J 16(4):e2000238. https://doi.org/10.1002/biot.202000238
11. Zhang Y-HP (2010) Production of biocommodities and bioelectricity by cell-free synthetic enzymatic pathway biotransformations: challenges and opportunities. Biotechnol Bioeng 105(4):663–677. https://doi.org/10.1002/bit.22630
12. Zhu Z, You C, Ma Y, Zhang Y-HPJ (2018) *In vitro* synthetic enzymatic biosystems at the interface of the food-energy-water nexus: a conceptual framework and recent advances. Process Biochem 74:43–49. https://doi.org/10.1016/j.procbio.2018.07.002
13. Zhu Z, Kin Tam T, Sun F, You C, Zhang Y-HP (2014) A high-energy-density sugar biobattery based on a synthetic enzymatic pathway. Nat Commun 5:3026. https://doi.org/10.1038/ncomms4026
14. Zhu Z, Zhang Y-HP (2017) *In vitro* metabolic engineering of bioelectricity generation by the complete oxidation of glucose. Metab Eng 39:110–116. https://doi.org/10.1016/j.ymben.2016.11.002

15. Wu R, Li F, Cui X, Li Z, Ma C, Jiang H et al (2023) Enzymatic electrosynthesis of glycine from CO_2 and NH_3. Angew Chem Int Ed Engl. https://doi.org/10.1002/ange.202218387
16. Li F, Wei X, Zhang L, Liu C, You C, Zhu Z (2022) Installing a green engine to drive an enzyme cascade: a light-powered *in vitro* biosystem for poly(3-hydroxybutyrate) synthesis. Angew Chem Int Ed Engl 61:e202111054. https://doi.org/10.1002/anie.202111054
17. Rollin JA, Tam TK, Zhang Y-HP (2013) New biotechnology paradigm: cell-free biosystems for biomanufacturing. Green Chem 15(7):1708–1719. https://doi.org/10.1039/C3GC40625C
18. Sun S, Wei X, You C (2019) The construction of an *in vitro* synthetic enzymatic biosystem that facilitates laminaribiose biosynthesis from maltodextrin and glucose. Biotechnol J 14(4): e1800493. https://doi.org/10.1002/biot.201800493
19. Li Y, Shi T, Han P, You C (2021) Thermodynamics-driven production of value-added D-allulose from inexpensive starch by an *in vitro* enzymatic synthetic biosystem. ACS Catal 11(9):5088–5099. https://doi.org/10.1021/acscatal.0c05718
20. Sun S, Wei X, Zhou X, You C (2021) Construction of an artificial *in vitro* synthetic enzymatic platform for upgrading low-cost starch to value-added disaccharides. J Agric Food Chem 69(1):302–314. https://doi.org/10.1021/acs.jafc.0c06936
21. You C, Shi T, Li YJ, Han PP, Zhou XG, Zhang Y-HP (2017) An *in vitro* synthetic biology platform for the industrial biomanufacturing of *myo*-inositol from starch. Biotechnol Bioeng 114(8):1855–1864. https://doi.org/10.1002/bit.26314
22. Meng D, Wei X, Zhang Y-HPJ, Zhu Z, You C, Ma Y (2018) Stoichiometric conversion of cellulosic biomass by *in vitro* synthetic enzymatic biosystems for biomanufacturing. ACS Catal 8(10):9550–9559. https://doi.org/10.1021/acscatal.8b02473
23. Wei X, Li Q, Hu C, You C (2021) An ATP-free *in vitro* synthetic enzymatic biosystem facilitating one-pot stoichiometric conversion of starch to mannitol. Appl Microbiol Biotechnol 105(5):1913–1924. https://doi.org/10.1007/s00253-021-11154-9
24. Schwander T, Schada von Borzyskowski L, Burgener S, Cortina NS, Erb TJ (2016) A synthetic pathway for the fixation of carbon dioxide *in vitro*. Science 354(6314):900–904. https://doi.org/10.1126/science.aah5237
25. Xie L, Wei X, Zhou X, Meng D, Zhou R, Zhang YPJ et al (2018) Conversion of D-glucose to L-lactate via pyruvate by an optimized cell-free enzymatic biosystem containing minimized reactions. Synth Syst Biotechnol 3(3):204–210. https://doi.org/10.1016/j.synbio.2018.05.003
26. Shi T, Liu S, Zhang YPJ (2019) CO_2 fixation for malate synthesis energized by starch via *in vitro* metabolic engineering. Metab Eng 55:152–160. https://doi.org/10.1016/j.ymben.2019.07.005
27. You C, Chen H, Myung S, Sathitsuksanoh N, Ma H, Zhang XZ et al (2013) Enzymatic transformation of nonfood biomass to starch. Proc Natl Acad Sci U S A 110(18):7182–7187. https://doi.org/10.1073/pnas.1302420110
28. Opgenorth PH, Korman TP, Bowie JU (2016) A synthetic biochemistry module for production of bio-based chemicals from glucose. Nat Chem Biol 12(6):393–395. https://doi.org/10.1038/nchembio.2062
29. Li Q, Ma Z, Meng D, Sui X, You C (2021) Facile biosynthesis of synthetic crystalline cellulose nanoribbon from maltodextrin through a minimized two-enzyme phosphorylase cascade and its application in emulsion. J Biotechnol 332:54–60. https://doi.org/10.1016/j.jbiotec.2021.03.018
30. Meng D, Wei X, Bai X, Zhou W, You C (2020) Artificial *in vitro* synthetic enzymatic biosystem for the one-pot sustainable biomanufacturing of glucosamine from starch and inorganic ammonia. ACS Catal 10(23):13809–13819. https://doi.org/10.1021/acscatal.0c03767
31. Zhu Z, Sun F, Zhang X, Zhang Y-HP (2012) Deep oxidation of glucose in enzymatic fuel cells through a synthetic enzymatic pathway containing a cascade of two thermostable dehydrogenases. Biosens Bioelectron 36(1):110–115. https://doi.org/10.1016/j.bios.2012.04.001

32. Wu R, Ma C, Zhang Y-HP, Zhu Z (2018) Complete oxidation of xylose for bioelectricity generation by reconstructing a bacterial xylose utilization pathway in vitro. ChemCatChem 10(9):2030–2035. https://doi.org/10.1002/cctc.201702018
33. Myung S, Rollin J, You C, Sun F, Chandrayan S, Adams MW et al (2014) *In vitro* metabolic engineering of hydrogen production at theoretical yield from sucrose. Metab Eng 24:70–77. https://doi.org/10.1016/j.ymben.2014.05.006
34. Rollin JA, Martin del Campo J, Myung S, Sun F, You C, Bakovic A et al (2015) High-yield hydrogen production from biomass by *in vitro* metabolic engineering: mixed sugars coutilization and kinetic modeling. Proc Natl Acad Sci U S A 112(16):4964–4969. https://doi.org/10.1073/pnas.1417719112
35. Moustafa HMA, Kim EJ, Zhu ZG, Wu CH, Zaghloul TI, Adams MWW et al (2016) Water splitting for high-yield hydrogen production energized by biomass xylooligosaccharides catalyzed by an enzyme cocktail. ChemCatChem 8(18):2898–2902. https://doi.org/10.1002/cctc.201600772
36. Kim EJ, Kim JE, Zhang Y-HPJ (2018) Ultra-rapid rates of water splitting for biohydrogen gas production through *in vitro* artificial enzymatic pathways. Energy. Environ Sci 11(8): 2064–2072. https://doi.org/10.1039/C8EE00774H
37. Wang W, Liu M, You C, Li Z, Zhang Y-HP (2017) ATP-free biosynthesis of a high-energy phosphate metabolite fructose 1,6-diphosphate by *in vitro* metabolic engineering. Metab Eng 42:168–174. https://doi.org/10.1016/j.ymben.2017.06.006
38. Wang W, Yang JG, Sun YX, Li ZM, You C (2020) Artificial ATP-free *in vitro* synthetic enzymatic biosystems facilitate aldolase-mediated C-C bond formation for biomanufacturing. ACS Catal 10(2):1264–1271. https://doi.org/10.1021/acscatal.9b04696
39. Cai T, Sun H, Qiao J, Zhu L, Zhang F, Zhang J et al (2021) Cell-free chemoenzymatic starch synthesis from carbon dioxide. Science 373(6562):1523–1527. https://doi.org/10.1126/science.abh4049
40. Flamholz A, Noor E, Bar-Even A, Milo R (2012) eQuilibrator – the biochemical thermodynamics calculator. Nucleic Acids Res 40:D770–D775. https://doi.org/10.1093/nar/gkr874
41. Beber ME, Gollub MG, Mozaffari D, Shebek KM, Flamholz AI, Milo R et al (2022) eQuilibrator 3.0: a database solution for thermodynamic constant estimation. Nucleic Acids Res 50(D1):D603–D6D9. https://doi.org/10.1093/nar/gkab1106
42. Meng D, Liang A, Wei X, You C (2019) Enzymatic characterization of a thermostable phosphatase from *Thermomicrobium roseum* and its application for biosynthesis of fructose from maltodextrin. Appl Microbiol Biotechnol 103(15):6129–6139. https://doi.org/10.1007/s00253-019-09917-6
43. Dai Y, Zhang J, Zhang T, Chen J, Hassanin HA, Jiang B (2020) Characteristics of a fructose 6-phosphate 4-epimerase from *Caldilinea aerophila* DSM 14535 and its application for biosynthesis of tagatose. Enzym Microb Technol 139:109594. https://doi.org/10.1016/j.enzmictec.2020.109594
44. Hu C, Wei X, Song Y (2022) A thermophilic phosphatase from *Methanothermobacter marburgensis* and its application to *in vitro* biosynthesis. Enzym Microb Technol 159: 110067. https://doi.org/10.1016/j.enzmictec.2022.110067
45. Tian C, Yang J, Li Y, Zhang T, Li J, Ren C et al (2020) Artificially designed routes for the conversion of starch to value-added mannosyl compounds through coupling *in vitro* and *in vivo* metabolic engineering strategies. Metab Eng 61:215–224. https://doi.org/10.1016/j.ymben.2020.06.008
46. Guterl JK, Garbe D, Carsten J, Steffler F, Sommer B, Reisse S et al (2012) Cell-free metabolic engineering: production of chemicals by minimized reaction cascades. ChemSusChem 5(11): 2165–2172. https://doi.org/10.1002/cssc.201200365
47. Opgenorth PH, Korman TP, Iancu L, Bowie JU (2017) A molecular rheostat maintains ATP levels to drive a synthetic biochemistry system. Nat Chem Biol 13(9):938–942. https://doi.org/10.1038/nchembio.2418

48. Bai X, Meng D, Wei X, Zhou X, Lu F, You C (2019) Facile synthesis of (−)-*vibo*-quercitol from maltodextrin via an *in vitro* synthetic enzymatic biosystem. Biotechnol Bioeng 116(10): 2710–2719. https://doi.org/10.1002/bit.27096
49. Satoh Y, Tajima K, Tannai H, Munekata M (2003) Enzyme-catalyzed poly (3-hydroxybutyrate) synthesis from acetate with CoA recycling and NADPH regeneration *in vitro*. J Biosci Bioeng 95(4):335–341
50. Xu ZN, Jing KJ, Liu Y, Cen PL (2007) High-level expression of recombinant glucose dehydrogenase and its application in NADPH regeneration. J Ind Microbiol Biotechnol 34(1):83–90. https://doi.org/10.1007/s10295-006-0168-2
51. Relyea HA, van der Donk WA (2005) Mechanism and applications of phosphite dehydrogenase. Bioorg Chem 33(3):171–189. https://doi.org/10.1016/j.bioorg.2005.01.003
52. Song YH, Liu MX, Xie LP, You C, Sun JS, Zhang Y-HPJ (2019) A recombinant 12-His tagged *Pyrococcus furiosus* soluble [NiFe]-hydrogenase I overexpressed in *Thermococcus kodakarensis* KOD1 facilitates hydrogen-powered *in vitro* NADH regeneration. Biotechnol J 14(4):e1800301. https://doi.org/10.1002/biot.201800301
53. Chen H, Zhang Y-HPJ (2021) Enzymatic regeneration and conservation of ATP: challenges and opportunities. Crit Rev Biotechnol 41(1):16–33. https://doi.org/10.1080/07388551.2020.1826403
54. Zhang X, Wu H, Huang B, Li ZM, Ye Q (2017) One-pot synthesis of glutathione by a two-enzyme cascade using a thermophilic ATP regeneration system. J Biotechnol 241:163–169. https://doi.org/10.1016/j.jbiotec.2016.11.034
55. Kamp AV, Klamt S (2020) MEMO: a method for computing metabolic modules for cell-free production systems. ACS Synth Biol 9(3):556–566. https://doi.org/10.1021/acssynbio.9b00434
56. Kim DM, Swartz JR (2001) Regeneration of adenosine triphosphate from glycolytic intermediates for cell-free protein synthesis. Biotechnol Bioeng 74:309–316. https://doi.org/10.1002/bit.1121
57. Wang Y, Zhang Y-HP (2009) Cell-free protein synthesis energized by slowly-metabolized maltodextrin. BMC Biotechnol 9(1):58. https://doi.org/10.1186/1472-6750-9-58
58. Wei X, Xie L, Zhang Y-HPJ, You C (2018) Stoichiometric regeneration of ATP by a NAD(P)/CoA-free and phosphate-balanced *in vitro* synthetic enzymatic biosystem. ChemCatChem 10(24):5597–5601. https://doi.org/10.1002/cctc.201801562
59. Song H, Ma C, Liu P, You C, Lin J, Zhu Z (2019) A hybrid CO_2 electroreduction system mediated by enzyme-cofactor conjugates coupled with Cu nanoparticle-catalyzed cofactor regeneration. J CO2 Util 34:568–575. https://doi.org/10.1016/j.jcou.2019.08.007
60. Zhang S, Shi J, Sun Y, Wu Y, Zhang Y, Cai Z et al (2019) Artificial thylakoid for the coordinated photoenzymatic reduction of carbon dioxide. ACS Catal 9(5):3913–3925. https://doi.org/10.1021/acscatal.9b00255
61. Fessner W-D, Walter C (1992) "Artificial metabolisms" for the asymmetric one-pot synthesis of branched-chain saccharides. Angew Chem Int Ed 31(5):614–616. https://doi.org/10.1002/anie.199206141
62. Zimmermann FT, Schneider A, Schörken U, Sprenger GA, Fessner W-D (1999) Efficient multi-enzymatic synthesis of D-xylulose 5-phosphate. Tetrahedron Asymmetry 10(9): 1643–1646. https://doi.org/10.1016/S0957-4166(99)00166-4
63. Wang L, Dash S, Ng CY, Maranas CD (2017) A review of computational tools for design and reconstruction of metabolic pathways. Synth Syst Biotechnol 2(4):243–252. https://doi.org/10.1016/j.synbio.2017.11.002
64. Li QK, Wang Z, Zhang M, Hou PF, Kang P (2017) Nitrogen doped tin oxide nanostructured catalysts for selective electrochemical reduction of carbon dioxide to formate. J Energy Chem 26(5):825–829. https://doi.org/10.1016/j.jechem.2017.08.010
65. Wang J, Li G, Li Z, Tang C, Feng Z, An H et al (2017) A highly selective and stable $ZnO\text{-}ZrO_2$ solid solution catalyst for CO_2 hydrogenation to methanol. Sci Adv 3(10):e1701290. https://doi.org/10.1126/sciadv.1701290

66. Ebrahim A, Lerman JA, Palsson BO, Hyduke DR (2013) COBRApy: COnstraints-based reconstruction and analysis for python. BMC Syst Biol 7:74. https://doi.org/10.1186/1752-0509-7-74
67. Yang X, Yuan Q, Luo H, Li F, Mao Y, Zhao X et al (2019) Systematic design and *in vitro* validation of novel one-carbon assimilation pathways. Metab Eng 56:142–153. https://doi.org/10.1016/j.ymben.2019.09.001
68. Caspi R, Billington R, Ferrer L, Foerster H, Fulcher CA, Keseler IM et al (2016) The MetaCyc database of metabolic pathways and enzymes and the BioCyc collection of pathway/genome databases. Nucleic Acids Res 44(D1):D471–D480. https://doi.org/10.1093/nar/gkv1164
69. Hadadi N, Hafner J, Shajkofci A, Zisaki A, Hatzimanikatis V (2016) ATLAS of biochemistry: a repository of all possible biochemical reactions for synthetic biology and metabolic engineering studies. ACS Synth Biol 5(10):1155–1166. https://doi.org/10.1021/acssynbio.6b00054
70. Jang WD, Kim GB, Kim Y, Lee SY (2022) Applications of artificial intelligence to enzyme and pathway design for metabolic engineering. Curr Opin Biotechnol 73:101–107. https://doi.org/10.1016/j.copbio.2021.07.024
71. Koch M, Duigou T, Faulon J-L (2020) Reinforcement learning for bioretrosynthesis. ACS Synth Biol 9(1):157–168. https://doi.org/10.1021/acssynbio.9b00447
72. Wei X, Meng D, You C (2020) *In vitro* metabolic engineering: current status and recent progress. In: Liu L, Du GC, Liu YF (eds) Systems and synthetic metabolic engineering. Elsevier, pp 183–206
73. Maini Rekdal V, Bess EN, Bisanz JE, Turnbaugh PJ, Balskus EP (2019) Discovery and inhibition of an interspecies gut bacterial pathway for Levodopa metabolism. Science 364(6445). https://doi.org/10.1126/science.aau6323
74. Li G, Ren J, Wu Q, Feng J, Zhu D, Ma Y (2013) Identification of a marine NADPH-dependent aldehyde reductase for chemoselective reduction of aldehydes. J Mol Catal B Enzym 90:17–22. https://doi.org/10.1016/j.molcatb.2013.01.010
75. Jia B, Han X, Kim KH, Jeon CO (2022) Discovery and mining of enzymes from the human gut microbiome. Trends Biotechnol 40(2):240–254. https://doi.org/10.1016/j.tibtech.2021.06.008
76. Lukk T, Sakai A, Kalyanaraman C, Brown SD, Imker HJ, Song L et al (2012) Homology models guide discovery of diverse enzyme specificities among dipeptide epimerases in the enolase superfamily. Proc Natl Acad Sci U S A 109(11):4122–4127. https://doi.org/10.1073/pnas.1112081109
77. Gerlt JA, Bouvier JT, Davidson DB, Imker HJ, Sadkhin B, Slater DR et al (2015) Enzyme function initiative-enzyme similarity tool (EFI-EST): a web tool for generating protein sequence similarity networks. Biochim Biophys Acta 1854(8):1019–1037. https://doi.org/10.1016/j.bbapap.2015.04.015
78. Burroughs AM, Allen KN, Dunaway-Mariano D, Aravind L (2006) Evolutionary genomics of the HAD superfamily: understanding the structural adaptations and catalytic diversity in a superfamily of phosphoesterases and allied enzymes. J Mol Biol 361(5):1003–1034. https://doi.org/10.1016/j.jmb.2006.06.049
79. Ren H, Shi C, Zhao H (2020) Computational tools for discovering and engineering natural product biosynthetic pathways. iScience 23(1):100795. https://doi.org/10.1016/j.isci.2019.100795
80. Yang G, Hong S, Yang P, Sun Y, Wang Y, Zhang P et al (2021) Discovery of an ene-reductase for initiating flavone and flavonol catabolism in gut bacteria. Nat Commun 12(1):790. https://doi.org/10.1038/s41467-021-20974-2
81. Ely B (2020) Recombination and gene loss occur simultaneously during bacterial horizontal gene transfer. PLoS One 15(1):e0227987. https://doi.org/10.1371/journal.pone.0227987
82. Mak WS, Tran S, Marcheschi R, Bertolani S, Thompson J, Baker D et al (2015) Integrative genomic mining for enzyme function to enable engineering of a non-natural biosynthetic pathway. Nat Commun 6:10005. https://doi.org/10.1038/ncomms10005

83. Song Y, DiMaio F, Wang RY, Kim D, Miles C, Brunette T et al (2013) High-resolution comparative modeling with RosettaCM. Structure 21(10):1735–1742. https://doi.org/10.1016/j.str.2013.08.005
84. Zhang Y, Skolnick J (2005) TM-align: a protein structure alignment algorithm based on the TM-score. Nucleic Acids Res 33(7):2302–2309. https://doi.org/10.1093/nar/gki524
85. Bonetta R, Valentino G (2020) Machine learning techniques for protein function prediction. Proteins: Struct, Funct, Bioinf 88(3):397–413. https://doi.org/10.1002/prot.25832
86. Dalkiran A, Rifaioglu AS, Martin MJ, Cetin-Atalay R, Atalay V, Doğan T (2018) ECPred: a tool for the prediction of the enzymatic functions of protein sequences based on the EC nomenclature. BMC Bioinformatics 19(1):334. https://doi.org/10.1186/s12859-018-2368-y
87. De Ferrari L, Mitchell JB (2014) From sequence to enzyme mechanism using multi-label machine learning. BMC Bioinformatics 15:150. https://doi.org/10.1186/1471-2105-15-150
88. Reetz MT, Sun Z, Ge Q (2023) Enzyme engineering: selective catalysts for applications in biotechnology, organic chemistry, and life science. Wiley, Weinheim
89. Singh RK, Tiwari MK, Singh R, Lee JK (2013) From protein engineering to immobilization: promising strategies for the upgrade of industrial enzymes. Int J Mol Sci 14(1):1232–1277. https://doi.org/10.3390/ijms14011232
90. Cedrone F, Menez A, Quemeneur E (2000) Tailoring new enzyme functions by rational redesign. Curr Opin Struct Biol 10(4):405–410. https://doi.org/10.1016/s0959-440x(00)00106-8
91. Waterhouse A, Bertoni M, Bienert S, Studer G, Tauriello G, Gumienny R et al (2018) SWISS-MODEL: homology modelling of protein structures and complexes. Nucleic Acids Res 46 (W1):W296–w303. https://doi.org/10.1093/nar/gky427
92. Yang J, Yan R, Roy A, Xu D, Poisson J, Zhang Y (2015) The I-TASSER suite: protein structure and function prediction. Nat Methods 12(1):7–8. https://doi.org/10.1038/nmeth.3213
93. Senior AW, Evans R, Jumper J, Kirkpatrick J, Sifre L, Green T et al (2020) Improved protein structure prediction using potentials from deep learning. Nature 577(7792):706–710. https://doi.org/10.1038/s41586-019-1923-7
94. Taylor RD, Jewsbury PJ, Essex JW (2002) A review of protein-small molecule docking methods. J Comput Aided Mol Des 16(3):151–166. https://doi.org/10.1023/a:1020155510718
95. Yang C, Chen EA, Zhang Y (2022) Protein-ligand docking in the machine-learning era. Molecules 27(14). https://doi.org/10.3390/molecules27144568
96. Tobe S, Shimogaki H, Ohdera M, Asai Y, Oba K, Iwama M et al (2006) Expression of *Bacillus* protease (Protease BYA) from *Bacillus sp.* Y in *Bacillus subtilis* and enhancement of its specific activity by site-directed mutagenesis-improvement in productivity of detergent enzyme. Biol Pharm Bull 29(1):26–33. https://doi.org/10.1248/bpb.29.26
97. Lee S, Lee DG, Jang MK, Jeon MJ, Jang HJ, Lee SH (2011) Improvement in the catalytic activity of β-agarase AgaA from *Zobellia galactanivorans* by site-directed mutagenesis. J Microbiol Biotechnol 21(11):1116–1122. https://doi.org/10.4014/jmb.1107.07001
98. Renugopalakrishnan V, Garduno-Juarez R, Narasimhan G, Verma CS, Wei X, Li P (2005) Rational design of thermally stable proteins: relevance to bionanotechnology. J Nanosci Nanotechnol 5(11):1759–1767. https://doi.org/10.1166/jnn.2005.441
99. Yu H, Yan Y, Zhang C, Dalby PA (2017) Two strategies to engineer flexible loops for improved enzyme thermostability. Sci Rep 7:41212. https://doi.org/10.1038/srep41212
100. Chuaboon L, Wongnate T, Punthong P, Kiattisewee C, Lawan N, Hsu CY et al (2019) One-pot bioconversion of L-arabinose to L-ribulose in an enzymatic cascade. Angew Chem Int Ed 58(8):2428–2432. https://doi.org/10.1002/anie.201814219
101. Roda S, Fernandez-Lopez L, Cañadas R, Santiago G, Ferrer M, Guallar V (2021) Computationally driven rational design of substrate promiscuity on serine ester hydrolases. ACS Catal 11(6):3590–3601. https://doi.org/10.1021/acscatal.0c05015
102. Chen H, Zhu ZG, Huang R, Zhang Y-HP (2016) Coenzyme engineering of a hyperthermophilic 6-phosphogluconate dehydrogenase from $NADP^+$ to NAD^+ with its application to biobatteries. Sci Rep 6:36311. https://doi.org/10.1038/srep36311

103. Xu LH, Du YL (2018) Rational and semi-rational engineering of cytochrome P450s for biotechnological applications. Synth Syst Biotechnol 3(4):283–290. https://doi.org/10.1016/j.synbio.2018.10.001
104. Chica RA, Doucet N, Pelletier JN (2005) Semi-rational approaches to engineering enzyme activity: combining the benefits of directed evolution and rational design. Curr Opin Biotechnol 16(4):378–384. https://doi.org/10.1016/j.copbio.2005.06.004
105. Voigt CA, Kauffman S, Wang ZG (2000) Rational evolutionary design: the theory of in vitro protein evolution. Adv Protein Chem 55:79–160. https://doi.org/10.1016/s0065-3233(01)55003-2
106. Reetz MT, Carballeira JD (2007) Iterative saturation mutagenesis (ISM) for rapid directed evolution of functional enzymes. Nat Protoc 2(4):891–903. https://doi.org/10.1038/nprot.2007.72
107. Reetz MT, Wu S (2008) Greatly reduced amino acid alphabets in directed evolution: making the right choice for saturation mutagenesis at homologous enzyme positions. Chem Commun (Camb) 43:5499–5501. https://doi.org/10.1039/b813388c
108. Dwyer MA, Looger LL, Hellinga HW (2004) Computational design of a biologically active enzyme. Science 304(5679):1967–1971. https://doi.org/10.1126/science.1098432
109. Reetz MT (2016) Selected examples of directed evolution of enzymes with emphasis on stereo- and regioselectivity, substrate scope, and/or activity. Directed evolution of selective enzymes: catalysts for organic chemistry and biotechnology, pp 167–203
110. Huang R, Chen H, Zhong C, Kim JE, Zhang YH (2016) High-throughput screening of coenzyme preference change of thermophilic 6-phosphogluconate dehydrogenase from $NADP^+$ to NAD^+. Sci Rep 6:32644. https://doi.org/10.1038/srep32644
111. Huang R, Chen H, Zhou W, Ma C, Zhang YP (2018) Engineering a thermostable highly active glucose 6-phosphate dehydrogenase and its application to hydrogen production *in vitro*. Appl Microbiol Biotechnol 102(7):3203–3215. https://doi.org/10.1007/s00253-018-8798-7
112. Ma C, Liu M, You C, Zhu Z (2020) Engineering a diaphorase via directed evolution for enzymatic biofuel cell application. Bioresour Bioprocess 7(1):23. https://doi.org/10.1186/s40643-020-00311-z
113. Zhou W, Huang R, Zhu ZG, Zhang Y-HPJ (2018) Coevolution of both thermostability and activity of polyphosphate glucokinase from *Thermobifida fusca* YX. Appl Environ Microbiol 84(16):e01224–e01218. https://doi.org/10.1128/AEM.01224-18
114. Yang KK, Wu Z, Arnold FH (2019) Machine-learning-guided directed evolution for protein engineering. Nat Methods 16(8):687–694. https://doi.org/10.1038/s41592-019-0496-6
115. Wittmann BJ, Johnston KE, Wu Z, Arnold FH (2021) Advances in machine learning for directed evolution. Curr Opin Struct Biol 69:11–18. https://doi.org/10.1016/j.sbi.2021.01.008
116. Fox RJ, Davis SC, Mundorff EC, Newman LM, Gavrilovic V, Ma SK et al (2007) Improving catalytic function by ProSAR-driven enzyme evolution. Nat Biotechnol 25(3):338–344. https://doi.org/10.1038/nbt1286
117. You C, Myung S, Zhang Y-HP (2012) Facilitated substrate channeling in a self-assembled trifunctional enzyme complex. Angew Chem Int Ed 51(35):8787–8790. https://doi.org/10.1002/anie.201202441
118. Han P, You C, Li Y, Shi T, Wu H, Zhang Y-HPJ (2023) High-titer production of *myo*-inositol by a co-immobilized four-enzyme cocktail in biomimetic mineralized microcapsules. Chem Eng J 461:141946. https://doi.org/10.1016/j.cej.2023.141946
119. Bauler P, Huber G, Leyh T, McCammon JA (2010) Channeling by proximity: the catalytic advantages of active site colocalization using Brownian dynamics. J Phys Chem Lett 1(9):1332–1335. https://doi.org/10.1021/jz1002007
120. Fan LW, Wang Y, Tuyishime P, Gao N, Li QG, Zheng P et al (2018) Engineering artificial fusion proteins for enhanced methanol bioconversion. ChemBioChem 19(23):2465–2471. https://doi.org/10.1002/cbic.201800424

121. Zakeri B, Fierer JO, Celik E, Chittock EC, Schwarz-Linek U, Moy VT et al (2012) Peptide tag forming a rapid covalent bond to a protein, through engineering a bacterial adhesin. Proc Natl Acad Sci U S A 109(12):E690–E697. https://doi.org/10.1073/pnas.1115485109
122. Fierobe HP, Mechaly A, Tardif C, Belaich A, Lamed R, Shoham Y et al (2001) Design and production of active cellulosome chimeras. Selective incorporation of dockerin-containing enzymes into defined functional complexes. J Biol Chem 276(24):21257–21261. https://doi.org/10.1074/jbc.M102082200
123. Fierobe HP, Bayer EA, Tardif C, Czjzek M, Mechaly A, Belaich A et al (2002) Degradation of cellulose substrates by cellulosome chimeras. Substrate targeting versus proximity of enzyme components. J Biol Chem 277(51):49621–49630. https://doi.org/10.1074/jbc.M207672200
124. Liu M, Song Y, Zhang YPJ, You C (2022) Carrier-free immobilization of multi-enzyme complex facilitates in vitro synthetic enzymatic biosystem for biomanufacturing. ChemSusChem:e202202153. https://doi.org/10.1002/cssc.202202153
125. Meng D, Wu R, Wang J, Zhu Z, You C (2019) Acceleration of cellodextrin phosphorolysis for bioelectricity generation from cellulosic biomass by integrating a synthetic two-enzyme complex into an *in vitro* synthetic enzymatic biosystem. Biotechnol Biofuels 12:267. https://doi.org/10.1186/s13068-019-1607-4
126. Chen H, Huang R, Kim E-J, Zhang Y-HPJ (2018) Building a thermostable metabolon for facilitating coenzyme transport and *in vitro* hydrogen production at elevated temperature. ChemSusChem 11(18):3120–3130. https://doi.org/10.1002/cssc.201801141
127. Sheldon RA, van Pelt S (2013) Enzyme immobilisation in biocatalysis: why, what and how. Chem Soc Rev 42(15):6223–6235. https://doi.org/10.1039/c3cs60075k
128. Han P, Zhou X, You C (2020) Efficient multi-enzymes immobilized on porous microspheres for producing inositol from starch. Front Bioeng Biotechnol:8. https://doi.org/10.3389/fbioe.2020.00380
129. Talekar S, Joshi A, Kambale S, Jadhav S, Nadar S, Ladole M (2017) A tri-enzyme magnetic nanobiocatalyst with one pot starch hydrolytic activity. Chem Eng J 325:80–90. https://doi.org/10.1016/j.cej.2017.05.054
130. Su HH, Guo ZW, Wu XL, Xu P, Li N, Zong MH et al (2019) Efficient bioconversion of sucrose to high-value-added glucaric acid by *in vitro* metabolic engineering. ChemSusChem 12(10):2278–2285. https://doi.org/10.1002/cssc.201900185
131. Li G, Wei X, Wu R, Zhou W, Li Y, Zhu Z et al (2022) Stoichiometric conversion of maltose for biomanufacturing by *in vitro* synthetic enzymatic biosystems. BioDesign Res 2022. https://doi.org/10.34133/2022/9806749
132. Zhong C, Wei P, Zhang Y-HP (2017) A kinetic model of one-pot rapid biotransformation of cellobiose from sucrose catalyzed by three thermophilic enzymes. Chem Eng Sci 161:159–166. https://doi.org/10.1016/j.ces.2016.11.047
133. Korman TP, Opgenorth PH, Bowie JU (2017) A synthetic biochemistry platform for cell free production of monoterpenes from glucose. Nat Commun 8:15526. https://doi.org/10.1038/ncomms15526
134. Takors R (2012) Scale-up of microbial processes: impacts, tools and open questions. J Biotechnol 160(1–2):3–9. https://doi.org/10.1016/j.jbiotec.2011.12.010
135. Dudley QM, Karim AS, Jewett MC (2015) Cell-free metabolic engineering: biomanufacturing beyond the cell. Biotechnol J 10(1):69–82. https://doi.org/10.1002/biot.201400330
136. Han P, Wang X, Li Y, Wu H, Shi T, Shi J (2023) Synthesis of a healthy sweetener D-tagatose from starch catalyzed by semiartificial cell factories. J Agric Food Chem 71(8):3813–3820. https://doi.org/10.1021/acs.jafc.2c08400
137. Xu X, Zhang W, You C, Fan C, Ji W, Park J-T et al (2023) Biosynthesis of artificial starch and microbial protein from agricultural residue. Sci Bull 68(2):214–223. https://doi.org/10.1016/j.scib.2023.01.006
138. Ye J, Li Y, Bai Y, Zhang T, Jiang W, Shi T et al (2022) A facile and robust T7-promoter-based high-expression of heterologous proteins in *Bacillus subtilis*. Bioresour Bioprocess 9(1):56. https://doi.org/10.1186/s40643-022-00540-4

139. Kim JE, Kim EJ, Chen H, Wu CH, Adams MWW, Zhang Y-HP (2017) Advanced water splitting for green hydrogen gas production through complete oxidation of starch by *in vitro* metabolic engineering. Metab Eng 44:246–252. https://doi.org/10.1016/j.ymben.2017.09.015
140. Lovelock SL, Crawshaw R, Basler S, Levy C, Baker D, Hilvert D et al (2022) The road to fully programmable protein catalysis. Nature 606(7912):49–58. https://doi.org/10.1038/s41586-022-04456-z
141. Yeh AH-W, Norn C, Kipnis Y, Tischer D, Pellock SJ, Evans D et al (2023) De novo design of luciferases using deep learning. Nature 614(7949):774–780. https://doi.org/10.1038/s41586-023-05696-3
142. Nikolados EM, Wongprommoon A, Aodha OM, Cambray G, Oyarzun DA (2022) Accuracy and data efficiency in deep learning models of protein expression. Nat Commun 13(1):7755. https://doi.org/10.1038/s41467-022-34902-5

Adv Biochem Eng Biotechnol (2023) 186: 29–50
https://doi.org/10.1007/10_2023_222

Published online: 13 June 2023

Cell-Free Production and Regeneration of Cofactors

Check for updates

Gladwin Suryatin Alim, Takuma Suzuki, and Kohsuke Honda

Contents

1 Introduction 30
2 Adenosine 5′-Triphosphate (ATP) 32
  2.1 Single-Enzyme Systems 32
  2.2 Multi-enzyme Cascade System 34
  2.3 ATP Regeneration from Adenosine (Ado) and Adenosine Monophosphate (AMP) . 35
  2.4 Considerations for Generating Other Nucleotides 36
3 Nicotinamide Adenine Dinucleotide (NAD(H)) and Nicotinamide Adenine Dinucleotide Phosphate (NADP(H)) 36
  3.1 Biosynthesis of NAD(H) and NADP(H) 37
  3.2 Chemo-Enzymatic Production of NAD(H) and NADP(H) 37
  3.3 Cell-Free Reconstitution of NAD(H) Salvage Pathway 39
  3.4 Regeneration of NADH and NADPH 39
4 Coenzyme A (CoA) 40
  4.1 Production of CoA 41
  4.2 Acylation and Deacylation of CoA 42
5 *S*-Adenosyl-L-Methionine (SAM) 43
  5.1 Enzymatic SAM Regeneration 44
  5.2 SAM Regeneration by Unnatural Methylating Agent 44
6 Concluding Remarks 45
References 46

G. Suryatin Alim
Department of Chemistry, University of Basel, Basel, Switzerland

International Center for Biotechnology, Osaka University, Osaka, Japan

T. Suzuki
International Center for Biotechnology, Osaka University, Osaka, Japan

K. Honda (✉)
International Center for Biotechnology, Osaka University, Osaka, Japan

Industrial Biotechnology Initiative Division, Institute for Open and Transdisciplinary Research Initiatives, Osaka University, Osaka, Japan
e-mail: honda.kohsuke.icb@osaka-u.ac.jp

Abstract Cofactors, such as adenosine triphosphate, nicotinamide adenine dinucleotide, and coenzyme A, are involved in nearly 50% of enzymatic reactions and widely used in biocatalytic production of useful chemicals. Although commercial production of cofactors has been mostly dependent on extraction from microbial cells, this approach has a theoretical limitation to achieve a high-titer, high-yield production of cofactors owing to the tight regulation of cofactor biosynthesis in living cells. Besides the cofactor production, their regeneration is also a key challenge to enable continuous use of costly cofactors and improve the feasibility of enzymatic chemical manufacturing. Construction and implementation of enzyme cascades for cofactor biosynthesis and regeneration in a cell-free environment can be a promising approach to these challenges. In this chapter, we present the available tools for cell-free cofactor production and regeneration, the pros and cons, and how they can contribute to promote the industrial application of enzymes.

Graphical Abstract

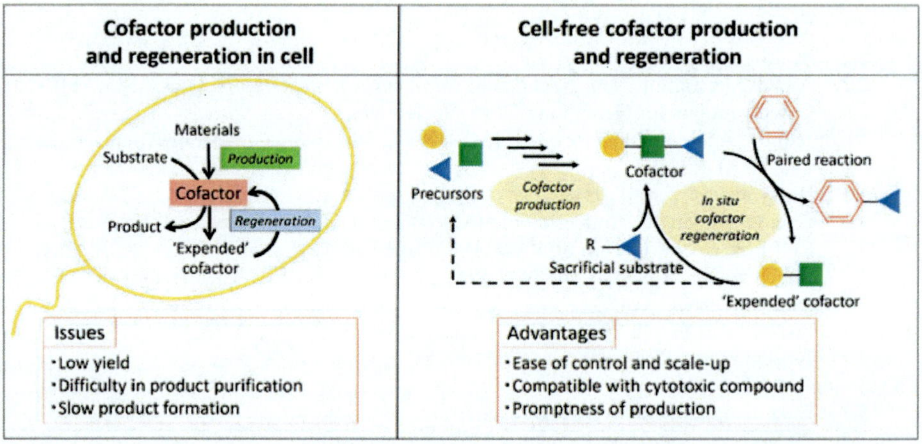

Keywords Adenosine triphosphate, Coenzyme A, Cofactor, Nicotinamide adenine dinucleotide (phosphate), *S*-adenosyl-L-methionine

1 Introduction

Enzymes are the molecular machinery that all organisms rely on for physiological functions such as metabolite uptake, energy metabolism, transport of molecules, and self-defense. As biocatalysts, enzymes work by lowering the activation energy of chemical reactions by introducing high-energy groups or by stabilizing the intermediates through intrinsic protein–ligand interactions. Unlike chemical catalysts, enzymes usually operate under mild conditions and they can be engineered to have broad or narrow substrate specificities and catalyze enantioselective reactions [1, 2]. Owing to their advantageous and unique properties, enzymes have been used for the production of food materials, drug molecules, detection of biomolecules for medical purposes, and as assisting reagents for washing detergents [3]. With the decreasing cost of enzyme production, an even higher demand for enzyme-related applications can be expected in the near future.

Enzymes usually work in combination with non-protein factors. Nearly 50% of all enzymes require non-protein factors for their catalytic activity [4]. These factors are classified into coenzymes and prosthetic groups, which are collectively called cofactors [4]. Common cofactors comprise adenosine 5′-triphosphate (ATP), which is required as a phosphate donor for ATP-dependent kinases, and nicotinamide adenine dinucleotide (NAD^+ and NADH, collectively referred to as NAD(H)), which is required as a donor and receptor of H^+ for dehydrogenases. However, cofactors are generally expensive depending on their production method, origin, and/or stability of the cofactors. For instance, commercial ATP is prepared from equine muscle, while NAD(H) is obtained by extraction from yeast (Sigma-Aldrich). Although cofactors can be obtained from living organisms in this manner, cofactors regulate numerous metabolic functions and are kept in homeostasis inside the cell [5]. As such, accumulating cofactors in vivo for the typical cell cultivation and extraction approach may be difficult and results in low yield. Not to mention, the cell extraction method faces issues such as slow product formation, cell toxicity, and difficulty in product purification [6]. Therefore, researchers are developing more efficient methods for cofactor production.

The cell-free production (or in vitro production) system is a promising technique because it has the following advantages: ease of control, ease of scale-up, compatibility with cytotoxic compounds, and promptness of production [7–9]. Owing to these advantages, cell-free production can be a good fit for cofactor production, avoiding the problem of cofactor production. Another advantage of cell-free cofactors production is the possibility to integrate it to another biocatalysis system requiring the cofactors. Cofactor-dependent enzymes that require a stoichiometric amount of cofactors for their function (e.g., ATP-dependent or NAD-dependent) usually suffer from substrate and product inhibition at high concentration [10]. Therefore, in situ cofactors regeneration that allows the biocatalytic system to operate at low concentrations of cofactors to avoid the inhibitions described above is often employed in biocatalysis systems.

In this chapter, we defined "cell-free cofactor production" as de novo production of cofactor from any substrates and "regeneration of cofactor" as restoring cofactor to its bioactive state after consumption. Here, we discuss the available tools for cell-free cofactor production and regeneration, the pros and cons, and how the field has evolved to date. From this chapter, we hope the reader can make an informed decision to which method should be applied for their specific application.

2 Adenosine 5′-Triphosphate (ATP)

ATP is an essential cofactor found in all domains of life. In enzymatic reactions, ATP is commonly used as an energy source to drive thermodynamically unfavorable reactions by transferring phosphate group to a substrate molecule. The expended ATP, usually in the form of ADP, is phosphorylated back to ATP by a series of catabolic pathways that differ according to the organism (Fig. 1a). According to the KEGG database, ATP is involved in over 500 enzymatic reactions, and as a key metabolite, ATP is generally required in many cell-free applications [11–14]. Thus, a robust ATP regeneration system is crucial for the advancement of cell-free production.

Bulk external addition of ATP can quickly become expensive ($50/g; 99% purity, A2383 Sigma Aldrich), and the accumulation of ADP or AMP can inhibit enzymes, especially kinases, in the reaction mixture owing to their intrinsic feedback regulation [15]. For these reasons, methods to replenish ATP in vitro, such that ATP is regenerated multiple times from ADP, have been developed. Current approaches to ATP regeneration issues differ with respect to the (1) substrate/energy source, (2) coenzyme requirement, (3) total number of enzymes involved, and (4) by-products. An ideal system should utilize cheap substrates, not require coenzymes, be easy to use (i.e., easy to purify and control), exhibit high stability, and produce non-disruptive by-products. Despite these advances, there is no perfect system in which these four criteria are met, and the proper choice has to be made based on the application.

In this chapter, we discuss the available tools for ATP regeneration from ADP, their advantages and disadvantages, and the advances made in recent years in pursuit of the creation of the best approach. With this as the basis, strategies for ATP generation from adenosine and AMP are discussed. Finally, options for expanding the application of ADP to ATP regeneration systems in the production of other nucleoside phosphates are discussed.

2.1 Single-Enzyme Systems

At a glance, a single-enzyme solution is preferable because it reduces the number of protein production-purification steps and is easier to control than a multiple-enzyme system (Fig. 1b). Creatine kinase (CK, EC 2.7.3.2) and pyruvate kinase (PK, EC 2.7.1.40) are two popular choices for laboratory-scale applications. They have high ATP regenerating activities, they are commercially available, and PK can be combined with lactate dehydrogenase for the regeneration of NADH [16]. However, the high cost and low stability of creatine phosphate and phosphoenolpyruvate hamper the use of CK and PK in higher-scale production. These two characteristics are general problems encountered when using phosphorylated molecules (e.g., acetyl phosphate, AcP) as phosphate donors.

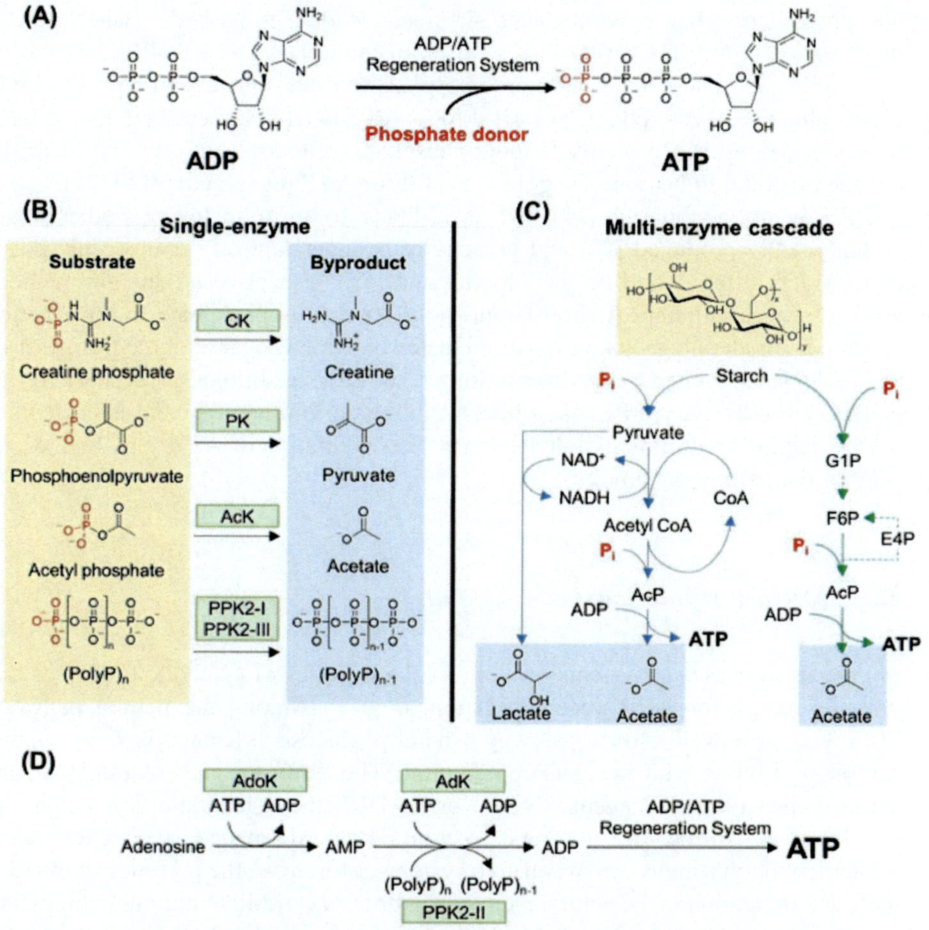

Fig. 1 Approaches for the production and regeneration of ATP. (**a**) General schematic illustration of ADP/ATP regeneration systems. (**b**) Common phosphate donors and their respective enzymes in single-enzyme reaction systems for ATP regeneration. (**c**) Multiple-enzyme cascade reactions with starch as the substrate for ADP/ATP regeneration. The NAD(H) and CoA-dependent pathway is indicated by the blue arrow. The cofactor-independent pathway is indicated by the green arrow. Only the main metabolites are shown in the figure. (**d**) Generation of ADP from adenosine or AMP. The enzymes are enclosed in green rectangles. Enzyme abbreviations: *CK* creatine kinase, *PK* pyruvate kinase, *AcK* acetate kinase, *PPK* polyphosphate kinase, *AdoK* adenosine kinase, *AdK* adenylate kinase. Compound abbreviations: *Pi* orthophosphate, *G1P* glucose 1-phosphate, *F6P* fructose 6-phosphate, *AcP* acetyl phosphate

Polyphosphate kinase (PPK, EC 2.7.4.1) catalyzes a reversible reaction between nucleoside phosphates and inorganic polyphosphate (polyP), and it has become a popular choice for ATP regeneration. PPK is grouped based on the reaction equilibrium, and the nucleoside phosphate substrate and product. The equilibrium of type 2 PPK (PPK2) is shifted toward the production of higher energy nucleoside phosphate rather than polyphosphate synthesis, leading to its wide usage for ATP regeneration [17–19]. Class II PPK2 (PPK2-II) and class I PPK2 (PPK2-I) specifically phosphorylate nucleoside monophosphate to diphosphate, and diphosphate to triphosphate, respectively. Class III PPK2 (PPKII-III) can catalyze the double phosphorylation from nucleoside monophosphate to its triphosphate, and PPK2-III has the potential to become the gold standard for ATP regeneration [20, 21].

PPK is not without drawbacks, as it tends to form inclusion bodies when recombinantly produced [20] and its activity is dependent on the inorganic polyP type, which differs based on the supplier and their respective production methods [19]. The accumulation of short-length polyP and orthophosphate in the reaction mixture is disadvantageous for repetitive batch reactions because of the formation of insoluble salts with divalent metal ions [22, 23]. Additionally, because of its generally low activity, PPK often becomes the rate-limiting step and is difficult to use in reactions requiring a high active pool concentration of ATP (e.g., high K_m to ATP of the partner enzyme).

2.2 *Multi-enzyme Cascade System*

Multi-enzyme cascade systems usually involve a series of catabolic processes that provide energy for ADP phosphorylation to ATP. Among the natural pathway, glycolysis is a well-known pathway whereby glucose is catabolized to smaller carbon molecules with net ATP production. The multi-enzyme requirement and use of expensive redox agents (NAD^+ or $NADP^+$, the regeneration of which will be discussed further in the next section) have dissuaded researchers from implementing this approach in cell-free systems. Moreover, the presence of intricate feedback regulation in the natural pathway hampers its usability, although this can be avoided by switching to feedback-insensitive enzymes [24]. On the other hand, the multi-enzyme approach can bypass the use of expensive high-energy sugar phosphates, instead starting with cheap, readily available metabolites. An early example of this system was a cascade comprising four enzymes powered by pyruvate with two cofactors, NAD^+ and CoA [25]. An extension to this system, which allows it to uptake maltodextrin and glucose, requires another additional 11 enzymes, bringing the total number of enzymes to 15 (Fig. 1c, blue arrow) [26]. Recently, ethanol has been used as feedstock in a cascade reaction that utilizes NAD^+ and CoA as cofactors for the production of isoprenol [27]. However, NAD^+ has low thermostability, and CoA has a tendency to form disulfide bonds with other thiol groups such as itself (see Sect. 3.3).

Non-oxidative glycolysis (NOG) is a synthetic pathway involving 12 enzymes that combines glycolysis with the pentose phosphate pathway. The key enzyme in this pathway is phosphoketolase, which catalyzes the production of acetyl phosphate from fructose 6-phosphate or xylulose 5-phosphate [28]. Combined with starch-degrading enzymes, NOG uses starch and orthophosphate as energy and phosphate sources, respectively, to regenerate ATP from ADP without the use of NAD^+ or CoA (Fig. 1c, green arrow) [29].

Despite the advantages of substrate usage, the multiple enzymes required for multi-enzyme systems hamper their practical usability and further testing potential. This issue can be addressed by constructing co-expression vectors for the simultaneous gene expression of enzymes in a single expression strain. Thus, all the necessary enzymes are expressed in a single host strain and can be purified simultaneously through affinity chromatography. Our group previously showed the capability of co-expression vectors for recombinant production of thermophilic enzymes, reaching higher yields compared to the use of individually expressed enzymes. Instead of affinity tags, using only thermophile-derived enzymes allowed the enzyme cocktail to be purified by simple heat treatment of cell lysates [30]. Combining the co-expression vector and thermophilic enzymes approach, ATP regeneration based on NOG (ArNOG) was reconstructed for one-step purification and proved its functionality for glycerol 3-phosphate production [31].

2.3 *ATP Regeneration from Adenosine (Ado) and Adenosine Monophosphate (AMP)*

As adenosine (Ado) is the nucleoside base of ATP, the generation of ATP can be initialized from Ado (Fig. 1d) [32]. The first phosphorylation of Ado to AMP is commonly performed by adenosine kinase (AdoK, EC 2.7.1.20), using nucleoside triphosphate (NTP) as the phosphate donor. AdoK generally has good substrate specificity for adenosine as the nucleoside receptor and accepts ATP as a phosphate donor. However, AdoK is almost exclusively found in eukaryotes, which may make it difficult to produce in *E. coli* recombinant systems owing to its lack of post-translational modification. Another disadvantage of AdoK is its intrinsic substrate inhibition at concentrations as low as 5 μM, and its K_m value toward ATP is in the millimolar range [33, 34]. Nucleoside kinase (NK; EC 2.7.1. B20) is the prokaryote equivalent of AdoK, albeit with a higher threshold for substrate inhibition, and as the name suggests, the substrate specificity of the nucleoside receptor and phosphate donor is broad [35].

Phosphorylation of AMP to ADP can be achieved using PPK2-II with polyP as the phosphate donor. If substrate specificity to adenosine base and high catalytic efficiency are required, adenylate kinase (AdK, EC 2.7.4.3) can be used as an alternative. *E. coli* AdK is highly thermostable and is thought to be a common contaminant that causes AMP accumulation in several cascade reactions where only ADP and ATP are present [24, 27, 36]. Under the correct conditions, where AMP is generated as a by-product, AdK rapidly catalyzes the phosphorylation of AMP owing to its low K_m value [37]. However, the requirement of AdK for ATP decreases the overall theoretical ATP availability and can decrease the yield of the cascade reaction.

2.4 *Considerations for Generating Other Nucleotides*

Enzymes specific to ATP are found throughout nature, perhaps because of their physiological function as the main phosphate donor. However, nucleoside kinases specific to cytidine, uridine, guanosine, or thymine are rarely found in nature. When these kinases are identified, regeneration pathways specific to the target nucleotide can be constructed and readily paired with ADP/ATP regeneration system. On the other hand, the promiscuous nature of nucleoside kinases (NK) allows it to be a generalist regenerating module for NTP [38]. The first phosphorylation of the nucleoside to its monophosphate is catalyzed by promiscuous NK. The second phosphorylation step is restricted and requires specific enzymes for each NMP: UMP-CMP kinase (EC 2.7.4.22) for UMP and CMP, guanylate kinase (EC 2.7.4.8) for GMP, and AdK for ATP. The final phosphorylation from NDP to NTP is catalyzed by nucleoside diphosphate kinase (NDK, EC 2.7.4.6). We note that some PPK2s have been reported to phosphorylate broad range of NMPs and it can potentially be used as single-enzyme solution for NTP regeneration [20].

3 Nicotinamide Adenine Dinucleotide (NAD(H)) and Nicotinamide Adenine Dinucleotide Phosphate (NADP(H))

Nicotinamide adenine dinucleotide (NAD^+), nicotinamide adenine dinucleotide phosphate ($NADP^+$), and the reduced molecules (NADH and NADPH) are ubiquitously distributed redox cofactors consisting of two chemically joined nucleotide molecules. NAD^+ and NADH (collectively referred to as NAD(H)) are typically used in catabolic pathways, including the Embden–Meyerhof–Parnas, citric acid cycle, and Leloir pathways, as well as in DNA repair [39–41]. $NADP^+$ and NADPH (NADP(H)) serve as the major cofactors in anabolism, such as photosynthesis and fatty acid biosynthesis [40, 42]. Thus, NAD(H) and NAD(P)H often play distinct roles in natural metabolism despite their minor structural differences at the 2′--position of the ribose ring [43]. It should also be noted, however, that the cofactor specificities of enzymes are not always restricted. For example, many alcohol dehydrogenases (EC 1.1.1.1, 1.1.1.2, and 1.1.1.71) can accept both NAD(H) and NADP(H) to exert their catalytic activity [44]. Eukaryotes have distinct types of isocitrate dehydrogenase (IDH), including NADP(H)-dependent (IDH1 and IDH2, EC 1.1.1.42) and NAD(H)-dependent ones (IDH3, EC 1.1.1.41) [45, 46]. Owing to their involvement in a variety of enzymatic reactions, NAD(H) and NADP(H) have a high commercial demand in some industrial sectors, such as chemical manufacturing, clinical diagnosis, and functional foods. Currently, the commercial production of these cofactors is highly dependent on extraction from yeast cells, but the development of more economically feasible processes is desired for their manufacture.

3.1 *Biosynthesis of NAD(H) and NADP(H)*

In most living cells, NAD(H) is synthesized through the kynurenine (de novo pathway), Preiss–Handler, and NAD^+ salvage pathways [47–51] (Fig. 2a). $NADP^+$ is synthesized by phosphorylation of NAD^+ mediated by NAD^+ kinases (NADK, EC 2.7.1.23). NADK deficiency has been demonstrated to have a lethal effect in many organisms, indicating the indispensability of this enzyme in NADP(H) biosynthesis [54, 55]. Tryptophan or aspartate serves as the starting material in the kynurenine pathway. They are converted to quinolinic acid by distinct sets of enzymes [48, 51], and then converted to NAD^+ by a three-step enzymatic reaction. The Preiss–Handler pathway utilizes nicotinic acid as the initial substrate and consists of three enzymatic reactions: nicotinic acid phosphoribosyltransferase (NAPRT, EC 6.3.4.21), nicotinate-nucleotide adenylyltransferase (NMNAT, EC 2.7.7.18), and NAD^+ synthase (NADS, EC 6.3.1.5) [56]. The NAD^+ salvage pathway recycles the nicotinamide generated by the spontaneous degradation of NAD^+. Nicotinamide is phosphoribosylated by nicotinamide phosphoribosyltransferase (EC 2.4.2.12), which is a similar but distinct enzyme from NAPRT, and is then converted into NAD^+ by NMNAT. In most prokaryotes, nicotinamide is hydrolyzed to form nicotinic acid by nicotinamidase (EC 3.5.1.19) and then converted back to NAD^+ via the Preiss–Handler pathway. Pinson et al. [57] demonstrated a significant correlation between intracellular concentrations of ATP and NAD^+, although it is unclear which of them drives this phenomenon, indicating that the biosyntheses of these two most important cofactors are synchronously regulated in living cells.

3.2 *Chemo-Enzymatic Production of NAD(H) and NADP(H)*

An early study of NAD(H) and NADP(H) production in a cell-free environment was reported by Walt et al. [52]. In this method, AMP is converted to NAD^+ and $NADP^+$ through a series of chemical and enzymatic reactions (Fig. 2b). Initially, nicotinamide mononucleotide (NMN), a direct precursor of NAD^+, was synthesized from AMP in a three-step chemical reaction with a molar yield of 25%. The resulting NMN was subjected to a reaction mediated by a co-immobilized enzyme cocktail composed of AcK, AdK, NAD^+ phosphorylase (EC 2.7.7.1), and pyrophosphatase (EC 3.6.1.1). With the addition of AcP for continuous ATP regeneration, NAD^+ (~15–20% purity) was obtained with a molar yield of 97% against NMN. NADK is used to phosphorylate NAD^+ to $NADP^+$. Because NAD^+ and $NADP^+$ are synthesized through enzymatic reactions, they are in a reaction mixture compatible with enzyme reactions and can be directly subjected to a variety of biocatalysts without isolation and purification.

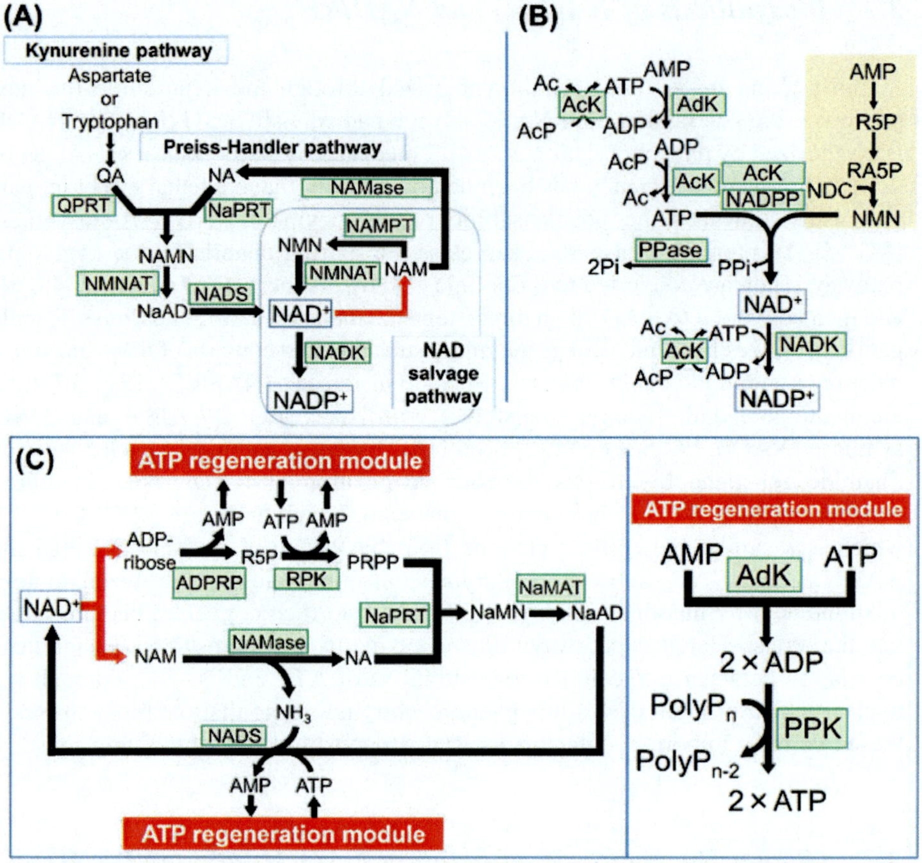

Fig. 2 NAD(P)$^+$ biosynthesis and approaches for cell-free NAD$^+$ production. (**a**) NAD(P)$^+$ biosynthesis. The red arrow represents spontaneous decomposition. (**b**) NAD(P)$^+$ production through a chemical and enzymatic cascade [52]. (**c**) Outline of the cell-free NAD$^+$ salvage pathway [53]. NAD$^+$ is recycled through six enzymatic reactions (black arrows) from ADP-ribose and nicotinamide generated by its thermal decomposition (red arrows). ATP is regenerated by a coupling reaction of AdK and PPK using polyP as a phosphate donor. Enzyme abbreviations: *QPRT* quinolinic acid phosphoribosyltransferase, *NaPRT* nicotinic acid phosphoribosyltranserase, *NAMase* nicotinamidase, *NMNAT* nicotinate-nucleotide adenylyltransferase, *NAMPT* nicotinamide phosphoribosyltransferase, *NADS* NAD$^+$ synthase, *NADK* NAD$^+$ kinase, *AcK* acetate kinase, *AdK* adenylate kinase, *NADPP* NAD$^+$ pyrophosphorylase, *PPase* pyrophosphatase, *ADPRP* ADP-ribose pyrophosphatase, *RPK* ribose-phosphate pyrophosphokinase, *PPK* pyrophosphokinase. Compound abbreviations: *QA* quinolinic acid, *NA* nicotinic acid, *NaMN* nicotinic acid mononucleotide, *NaAD* nicotinic acid adenine dinucleotide, *NAM* nicotinamide, *NMN* nicotinamide mononucleotide, *Ac* acetate, *AcP* acetyl phosphate, *NDC* 1-(2,4-dinitrophenyl)-3-carbamoylpyridium chloride, *Pi* orthophosphate, *R5P* ribose 5-phosphate, *RA5P* ribosylamine 5-phosphate, *PRPP* phosphoribosyl pyrophosphate

3.3 Cell-Free Reconstitution of NAD(H) Salvage Pathway

One of the limitations of the industrial use of NAD(H) and NADP(H) is their chemical instability, particularly at high temperatures. A possible solution to this problem is the use of chemically synthesized, thermally stable biomimetic cofactors, such as carba-$NADP^+$ [58]. However, as the acceptance of these biomimetics by natural enzymes is not high, a protein-engineering effort is needed to improve the affinity of the enzymes for these artificial cofactors. Our group addressed this issue by reconstituting the NAD(H) salvage pathway in a cell-free environment using recombinantly produced thermophilic enzymes [37, 53]. Through a six-step enzymatic reaction, NAD^+ can be produced from its decomposition products, ADP-ribose and nicotinamide (Fig. 2c). The ATP required to drive the pathway was regenerated from AMP using an enzyme set of AdK and PPK (see Sect. 2.3). In the presence of the reconstituted pathway, the NAD^+ concentration was almost constant for 15 h at 60°C, whereas more than 75% of the cofactor was decomposed in the control experiment without the pathway [53].

3.4 Regeneration of NADH and NADPH

As the maintenance of the intracellular balance between NAD^+/NADH and $NADP^+$/NADPH is a crucial issue in living cells, cells are equipped with regulatory machinery to maintain cofactor homeostasis. However, due to the absence of such molecular machineries in cell-free systems, it is indispensable to integrate an appropriate redox-cofactor-regeneration system with cell-free enzymatic reactions, such as those for chemical manufacturing, to avoid stoichiometric use of these expensive cofactors. NAD(P)H-dependent oxidoreductases are often used in the enzymatic production of enantiomerically pure chemicals on an industrial scale. Accordingly, a variety of NAD(P)H regeneration systems have been developed to date using chemical, electrochemical, photochemical, and enzymatic approaches [59–61]. Among these, enzymatic approaches are the most widely employed owing to their compatibility with aqueous conditions, where biocatalytic chemical manufacturing is usually implemented. Enzymatic NAD(P)H regeneration can be classified into four types [59, 62]. The use of a second enzyme and substrate is the most frequently used approach. NAD^+ and $NADP^+$ released by a reaction of interest are reduced back to NADH and NADPH, respectively, through the oxidation of cheap, sacrificial co-substrates catalyzed by a second enzyme (Fig. 3a). Glucose dehydrogenase (EC 1.1.1.118) and formate dehydrogenase (EC 1.17.1.9) are often employed as secondary enzymes because their reaction products (i.e., gluconolactone and carbon dioxide) can be readily removed from the reaction mixture through spontaneous hydrolysis and vaporization, and the equilibrium of the overall reaction can be maintained in the forward direction. A single enzyme with a second substrate system can be used when the enzyme catalyzing the main reaction (NADH- or NADPH-

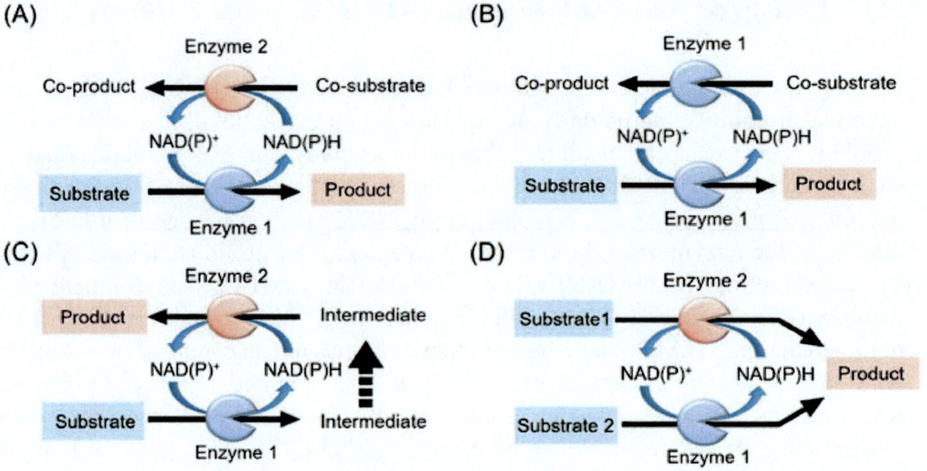

Fig. 3 Cell-free NAD(P)H regeneration systems. (**a**) Second enzyme with second substrate system, (**b**) single enzyme with second substrate system, (**c**) closed-loop system, and (**d**) convergent system

dependent reduction) can also catalyze the oxidation of sacrificial co-substrates (Fig. 3b). A typical example is the production of enantiomerically pure secondary alcohols from prochiral ketones using alcohol dehydrogenases, in which isopropanol is used as a co-substrate. The oxidation of isopropanol produces acetone, which can be readily made to evaporate from the reaction mixture and thereby maintains the reaction equilibrium. A closed-loop system has also been reported as an alternative NAD(H)/ NADP(H) regeneration approach. In this system, an enzyme oxidizes (or reduces) a substrate to a reaction intermediate using NAD(P)$^+$ (or NAD(P)H), followed by another enzymatic reaction catalyzing the reduction (or oxidation) of the intermediate to form the target product with NAD(P)H (or NAD(P)$^+$) (Fig. 3c). The convergent system implements NAD(P)$^+$-dependent oxidation and NAD(P)H-dependent reduction of two distinct substrates in a single batch, but both reactions yield an identical product (Fig. 3d). Using this system, Bornadel et al. [62] demonstrated ε-caprolactone production from cyclohexanone and 1,6-hexanediol using NAD(P)H-dependent Baeyer–Villiger monooxygenase (EC 1.14.13.–) and NAD (P)$^+$-dependent alcohol dehydrogenase.

4 Coenzyme A (CoA)

Coenzyme A (CoA) is an essential cofactor present in all domains of life and is mainly used as an acyl group transfer agent in numerous metabolic pathways, including fatty acid metabolism, pyruvate oxidation through the tricarboxylic acid (TCA) cycle, and the production of secondary metabolites [63, 64]. Acyl groups are fused to the free thiol group of CoA to produce CoA thioesters, typically in the form of acetyl-CoA. The activated carboxyl groups can act as an electrophile and a nucleophile at the C1 and C2 positions, respectively. Thus, CoA acts as a shuttle molecule for the transfer of acyl groups to the target compound.

4.1 Production of CoA

CoA is generally stable in buffers and does not degrade over long periods of incubation. Moreover, CoA is usually not consumed in the reaction and only acts as a shuttle molecule. Thus, only a single addition of a catalytic amount of CoA is required, and regeneration focuses on the acylation of CoA to acyl-CoA. Nonetheless, at the time of writing, commercial CoA is sold for \$3,000/g (≥93% purity, C3019 Sigma-Aldrich) and the prices are higher for the thioesters. With acetyl-CoA costing \$13,800/g (≥93% purity, A2181 Sigma-Aldrich) and malonyl-CoA \$18,200/g (>90% purity, M4263 Sigma-Aldrich). The price can quickly add up for large-scale reactions. We believe that it is beneficial for the catalytic amount of CoA to be produced in the same reaction vessel as the intended cascade pathway.

Techniques for cell-free in situ production of CoA have been developed over the years through total synthesis, chemoenzymatic, or fully enzymatic processes [24, 65, 66]. The biosynthesis of CoA in many organisms starts with D-pantothenate (or vitamin B5) and is converted to CoA after ligation of cysteine at the carboxyl end and adenosine phosphates at the hydroxyl end. The synthetic biomimetic pathway was selected for D-pantetheine as the starting substrate instead of D-pantothenate (Fig. 4a). D-Pantetheine is a D-pantothenate derivative with a cysteine moiety already attached to the carboxyl end, and its disulfide form is sold commercially at a relatively low price of \$100/g (>90% purity, P2125 Sigma-Aldrich). The first enzyme kinase in the cascade, PanK (EC 2.7.1.33), has promiscuous activity and accepts D-pantetheine as an alternative substrate to produce 4-phosphopantetheine (4-PP), the substrate for PPAT (EC 2.7.7.3). This decreases the number of enzymes and avoids the use of costly CTP required for cysteine ligation. Then, adenosine is transferred to 4-PP by PPAT, and a final phosphorylation at the 3′ position of adenosine by DPCK (EC 2.7.1.24) results in the generation of CoA.

In the one-pot batch reaction for CoA production, our group has demonstrated the production of 0.93 mM CoA (47% yield from D-pantetheine) in under 6 h by thermophilic enzyme catalysis. These enzymes were heterologously produced in *E. coli* and underwent only heat purification. Additionally, the reaction was performed without a complementary ATP regeneration pathway, and we later found that the systems were inhibited by ADP by-products. We expect that the yield and rate can be improved by implementing ATP regeneration and optimizing reaction conditions. Similarly, Mouterde and Stewart demonstrated a fully enzymatic fed-batch reaction using the same set of column-purified enzymes originated from *E. coli* [65]. They reported an almost complete conversion from D-pantetheine

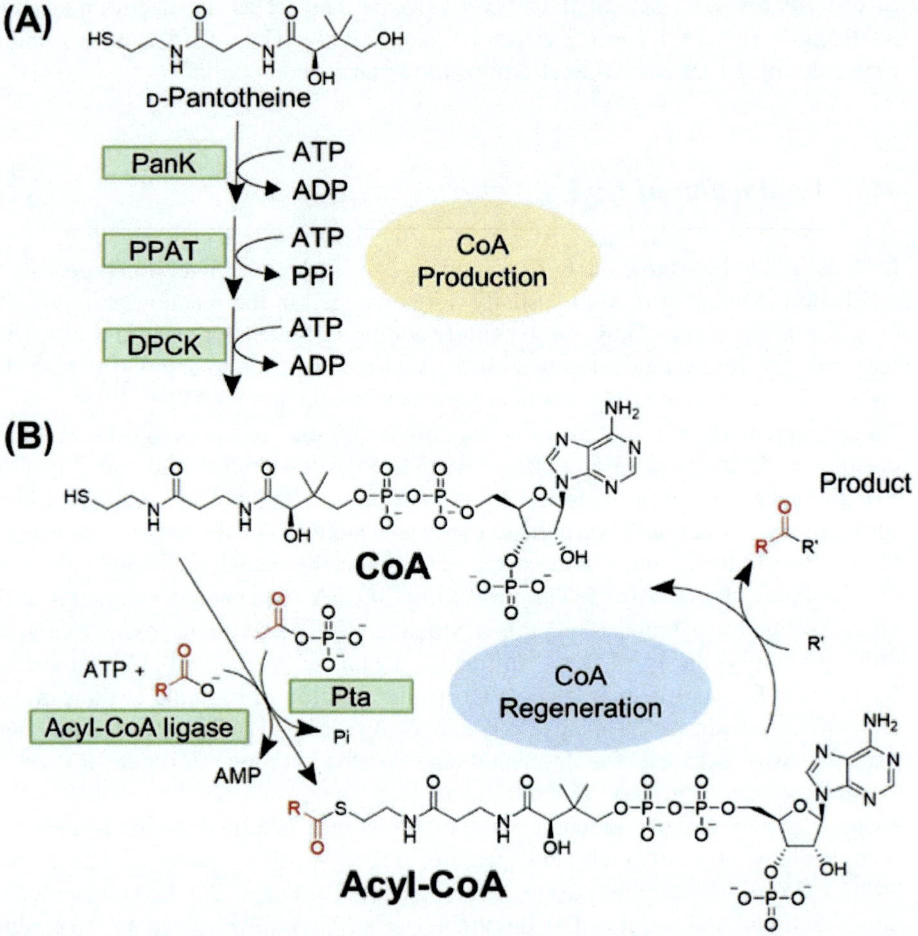

Fig. 4 Schematic illustration for the enzymatic production and regeneration of CoA. (**a**) Production of CoA from D-pantetheine via a three-step enzymatic reaction. (**b**) Regeneration of acyl-CoA from CoA by acyltransferases. The enzymes are enclosed in green rectangles. Abbreviations: *PanK* pantothenate kinase, *PPAT* phosphopantetheine adenylyltransferase, *DPCK* dephospho-CoA kinase, *Pta* phosphotransacetylase, *PPi* pyrophosphate

to CoA, although more steps are required in their approach. Each enzyme and ATP were added sequentially, and the reaction took more than 2 days to complete.

4.2 Acylation and Deacylation of CoA

The enzymatic formation of acyl-CoA from CoA generally requires a single enzyme. Two of the most commonly used enzymes are phosphotransacetylase (Pta; EC 2.3.1.8) and acyl-CoA ligase (or synthetase; EC 6.2.1.–) (Fig. 4b) [67]. Pta functions with acetyl phosphate (AcP) and does not require ATP as an energy supplier. AcP is easy to prepare and generally inexpensive [68], although, as discussed in the ATP section, it is quite unstable and should be generated in situ instead of external addition. On the other hand, acyl-CoA ligases function by activating the carboxylate of acyl donors through adenylation. In contrast to Pta, which uses high-energy AcP, acetyl-CoA ligase (ACL; EC 6.2.1.1) produces acetyl-CoA from acetate and ATP. Although some protists and archaea are known to possess ACLs that catalyze ADP-forming reactions (EC 6.2.1.13), typical ACLs produce AMP as the by-product of CoA acylation. From an energetic viewpoint, Pta is more efficient than ACL, as AcP can be regenerated using ATP by AcK, releasing ADP as the by-product. However, Pta can only be used to generate acetyl-CoA, whereas many acyl-CoA ligases can accept acyl groups of different lengths and produce varying CoA thioesters. A common example is malonyl-CoA ligase (MatB; EC 6.2.1.76), which has been used as an extender unit to generate diverse polyketide products [69–71]. A more specialized example is 4-coumarate-CoA ligase (EC 6.2.1.12) for the production of rosmarinic acid from caffeic acid [72]. Other options to regenerate acyl-CoA include coupling with a redox reaction, such as in the case of acetaldehyde dehydrogenase (EC 1.2.1.10) [27].

5 *S*-Adenosyl-L-Methionine (SAM)

Methylation is a common reaction in natural product biosynthesis and signal transduction [73]. The introduction of methyl groups to biomolecules can significantly change their physicochemical properties and create better therapeutic compounds [74]. The majority of these methylations are mediated by *S*-adenosyl-L-methionine (SAM)-dependent methyltransferases (MTs). SAM is very costly (≥$10,000/g, ≥80% purity, A4377 Sigma Aldrich) and generally labile because of its highly electrophilic sulfonium sulfur. Moreover, SAM has two chiral centers at the carbon α amino position and in the sulfonium sulfur and only the (*S,S*)-form is the biologically active form. Owing to spontaneous conversion racemization, commercially available SAM generally contains a considerable amount of inactive (*R,S*)-form [75]. Generating SAM in situ for cell-free chemical production is favorable in terms of both cost and stability. Most SAMs are used in a methyltransferase reaction by SAM-dependent MTs, which generate *S*-adenosyl-L-homocysteine (SAH) as a by-product. Another class of SAM-dependent enzymes, notably radical SAM enzymes, generates 5′-deoxyadenosine instead [76]. Since the biotechnological application of radical SAM is currently limited, this report describes the regeneration of SAM from SAH.

5.1 Enzymatic SAM Regeneration

Current approaches to SAM regeneration differentiate between the installation and/or enhancement of natural and artificial SAM biosynthesis pathways in the cell, or the use of unnatural compounds for direct SAM regeneration [77]. SAM is produced in vivo by L-methionine adenosyltransferase (MATs, EC 2.5.1.6) from ATP and L-methionine (Fig. 5, blue arrow). After transferring its methyl group in an S_N2-type reaction, the demethylated product, *S*-adenosylhomocysteine (SAH), is hydrolyzed by SAH hydrolase (SAHH, EC 3.13.2.1) to produce adenosine and L-Homocysteine. Adenosine is phosphorylated three times to ATP by a set of kinases, as described in the ATP regeneration section, and is ready to be used by MATs to generate SAM. Since L-methionine is inexpensive, external addition is feasible, and regeneration from L-homocysteine is not particularly necessary. However, inhibition by free thiol L-homocysteine needs to be considered to optimize the system. L-Homocysteine can be methylated back to L-methionine by L-homocysteine *S*-methyltransferase (HSMT, EC 2.1.1.10), using L-methylmethionine as the methyl donor. The bicyclic adenosine and L-homocysteine recycling for the SAM regeneration system reached up to 200 turnovers [32]. We note that L-methylmethionine was not available from Sigma-Aldrich at the time of writing and may need to be synthesized in the laboratory.

5.2 SAM Regeneration by Unnatural Methylating Agent

Single-enzyme SAM regeneration was spearheaded by the discovery of the ability of halide methyltransferase (HMT, EC 2.1.1.165) to methylate SAH in the presence of a strong electrophilic methylating reagent, such as methyl iodide (MeI) (Fig. 5, orange arrow) [78]. This system generally works in a cascade reaction with various MTs with turnover numbers of hundreds. The main drawback is the volatility (boiling point 42°C) and toxicity of MeI, which limits its use under mild and low-temperature conditions.

Recent developments in this field have focused on the use of less volatile methylating reagents and testing of different classes of MTs in the thiopurine methyltransferases (TPMTs) family [79, 80]. Instead of MeI, common methylating agents, such as dimethyl sulfate and methyl toluene sulfonate (MeOTs), are accepted by TPMTs as methylating agents for SAM regeneration. The stability of MeOTs is much greater than that of MeI, with a half-life in phosphate buffer of approximately 26 h and a boiling point of 292°C [80]. As this technique progresses, the development of a more stable and nontoxic methyl donor is an important step toward its industrial implementation.

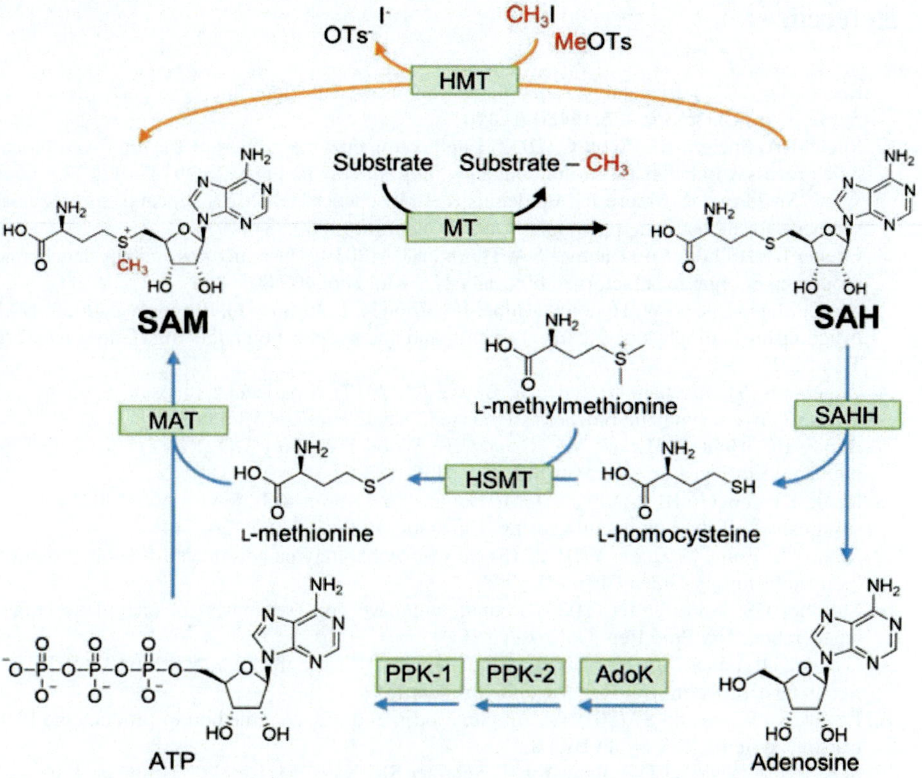

Fig. 5 Schematic illustration of the regeneration of SAM from SAH. The biomimetic enzymatic pathways are indicated by the blue arrows (bottom). The regeneration by unnatural substrate facilitated by HMT is shown by the orange arrow (top). Abbreviations: *SAHH* SAH hydrolase, *HSMT* L-homocysteine *S*-methyltransferase, *MAT* L-methionine adenosyltransferase, *HMT* halide methyltransferase, *AdoK* adenosine kinase, *PPK* polyphosphate kinase

6 Concluding Remarks

With the ever-increasing application of enzymes as "green" catalysts in various biotechnological fields, the importance of generating their respective cofactors cannot be overstated. In this chapter, we describe cell-free production and regeneration of biotechnologically relevant cofactors. Owing to their pivotal roles in natural metabolism, the intracellular pool sizes of cofactors are tightly regulated by diverse molecular machineries, indicating that there is a theoretical limitation in the improvement of cofactor manufacturing with conventional cell extraction approaches. Cell-free approaches can potentially overcome this limitation and enable the development of regulation-free and easily controllable manufacturing processes for cofactors and their derivatives. Despite their ubiquitous nature, there are significant variations in the pathways and enzymes involved in cofactor biosynthesis. In addition to process development, the exploration of novel and unique cofactor biosynthetic enzymes is a key challenge to the development of more feasible synthetic routes for cell-free cofactor production.

References

1. Reetz MT (2013) Biocatalysis in organic chemistry and biotechnology: past, present, and future. J Am Chem Soc 135:12480–12496
2. Rocha RA, Speight RE, Scott C (2022) Engineering enzyme properties for improved biocatalytic processes in batch and continuous flow. Org Process Res Dev 26:1914–1924
3. Wu S, Snajdrova R, Moore JC, Baldenius K, Bornscheuer UT (2020) Biocatalysis: enzymatic synthesis for industrial applications. Angew Chem Int Ed 60:88–119
4. Fischer JD, Holliday GL, Rahman SA, Thornton JM (2010) The structures and physicochemical properties of organic cofactors in biocatalysis. J Mol Biol 403:803–824
5. Xie N, Zhang L, Gao W, Huang C, Huber PE, Zhou X, Li C, Shen G, Bingwen Z (2020) NAD^+ metabolism: pathophysiologic mechanisms and therapeutic potential. Sig Transduct Target Ther 5:227
6. Opgenorth PH, Korman TP, Iancu L, Bowie JU (2017) A molecular rheostat maintains ATP levels to drive a synthetic biochemistry system. Nat Chem Biol 13:938–942
7. Bundy BC, Hunt JP, Jewett MC, Swartz JR, Wood DW, Frey DD, Rao G (2018) Cell-free biomanufacturing. Curr Opin Chem Eng 22:177–183
8. Rasor BJ, Vögeli B, Landwehr GM, Bogart JW, Karim AS, Jewett MC (2021) Toward sustainable, cell-free biomanufacturing. Curr Opin Biotechnol 69:136–144
9. Rollin JA, Tam TK, Zhang YHP (2013) New biotechnology paradigm: cell-free biosystems for biomanufacturing. Green Chem 15:1708
10. Mordhorst S, Andexer JN (2020) Round, round we go – strategies for enzymatic cofactor regeneration. Nat Prod Rep 37:1316–1333
11. Garenne D, Haines MC, Romantseva EF, Freemont P, Strychalski EA, Noireaux V (2021) Cell-free gene expression. Nat Rev Methods Primers 1:49
12. Lavickova B, Maerkl SJ (2019) A simple, robust, and low-cost method to produce the PURE cell-free system. ACS Synth Biol 8:455–462
13. Scheffen M, Marchal DG, Beneyton T, Schuller SK, Klose M, Diehl C, Lehmann J, Pfister P, Carrillo M, He H et al (2021) A new-to-nature carboxylation module to improve natural and synthetic CO2 fixation. Nat Catal 4:105–115
14. Wei F, Yuan R, Wen Q, Wen L (2023) Systematic enzymatic synthesis of dTDP-activated sugar nucleotides. Angew Chem Int Ed:e202217894
15. Gais HJ, Theil F (2002) Enzyme catalysis in organic synthesis. Wiley, pp 335–578
16. Luo S, Lin PP, Nieh LY, Liao GB, Tang PW, Chen C, Liao JC (2022) A cell-free self-replenishing CO_2-fixing system. Nat Catal 5:154–162
17. Cai T, Sun H, Qiao J, Zhu L, Zhang F, Zhang J, Tang Z, Wei X, Yang J, Yuan Q, Wang W, Yang X, Chu H, Wang Q, You C, Ma H, Sun Y, Li Y, Li C, Jiang H, Wang Q, Ma Y (2021) Cell-free chemoenzymatic starch synthesis from carbon dioxide. Science 373:1523–1527
18. Hanatani Y, Imura M, Taniguchi H, Okano K, Toya Y, Iwakiri R, Honda K (2019) *In vitro* production of cysteine from glucose. Appl Microbiol Biotechnol 103:8009–8019
19. Restiawaty E, Iwasa Y, Maya S, Honda K, Omasa T, Hirota R, Kuroda A, Ohtake H (2011) Feasibility of thermophilic adenosine triphosphate-regeneration system using *Thermus thermophilus* polyphosphate kinase. Process Biochem 46:1747–1752
20. Motomura K, Hirota R, Okada M, Ikeda T, Ishida T, Kuroda A (2014) A new subfamily of polyphosphate kinase 2 (class III PPK2) catalyzes both nucleoside monophosphate phosphorylation and nucleoside diphosphate phosphorylation. Appl Environ Microbiol 80:2602–2608
21. Ogawa M, Uyeda A, Harada K, Sato Y, Kato Y, Watanabe H, Honda K, Matsuura T (2019) Class III polyphosphate kinase 2 enzymes catalyze the pyrophosphorylation of adenosine-5′-monophosphate. Chembiochem 20:2961–2967
22. Fischöder T, Wahl C, Zerhusen C, Elling L (2019) Repetitive batch mode facilitates enzymatic synthesis of the nucleotide sugars UDP-gal, UDP-GlcNAc, and UDP-GalNAc on a multi-gram scale. Biotechnol J 14:1800386

23. Gottschalk J, Aßmann M, Kuballa J, Elling L (2022) Repetitive synthesis of high-molecular-weight hyaluronic acid with immobilized enzyme cascades. ChemSusChem 15:e202101071
24. Suryatin Alim G, Iwatani T, Okano K, Kitani S, Honda K (2021) *In vitro* production of coenzyme A using thermophilic enzymes. Appl Environ Microbiol 87:e0054121
25. Kim DM, Swartz JR (2001) Regeneration of adenosine triphosphate from glycolytic intermediates for cell-free protein synthesis. Biotechnol Bioeng 74:309–316
26. Wang Y, Zhang YHP (2009) Cell-free protein synthesis energized by slowly-metabolized maltodextrin. BMC Biotechnol 9:58
27. Liu H, Arbing MA, Bowie JU (2022) Expanding the use of ethanol as a feedstock for cell-free synthetic biochemistry by implementing acetyl-CoA and ATP generating pathways. Sci Rep 12: 7700
28. Bogorad IW, Lin T-S, Liao JC (2013) Synthetic non-oxidative glycolysis enables complete carbon conservation. Nature 502:693–697
29. Wei X, Xie L, Zhang YHP, You C (2018) Stoichiometric regeneration of ATP by a NAD(P)/ CoA-free and phosphate-balanced *in vitro* synthetic enzymatic biosystem. ChemCatChem 10: 5597–5601
30. Ninh PH, Honda K, Sakai T, Okano K, Ohtake H (2015) Assembly and multiple gene expression of thermophilic enzymes in *Escherichia coli* for *in vitro* metabolic engineering. Biotechnol Bioeng 112:189–196
31. Suryatin Alim G, Okano K, Honda K (2022) One-step preparation of cell-free ATP regeneration module based on non-oxidative glycolysis using thermophilic enzymes. Chembiochem 23: e202200210
32. Popadić D, Mhaindarkar D, Thai MHND, Hailes HC, Mordhorst S, Andexer JN (2021) A bicyclic *S*-adenosylmethionine regeneration system applicable with different nucleosides or nucleotides as cofactor building blocks. RSC Chem Biol 2:883–891
33. Barrado P, Rodríguez MJ, Jiménez A, Lobato MF (2003) Expression in *Escherichia coli* of a recombinant adenosine kinase from *Saccharomyces cerevisiae*: purification, kinetics and substrate analyses. Yeast 20:1145–1150
34. Long MC, Escuyer V, Parker WB (2003) Identification and characterization of a unique adenosine kinase from *Mycobacterium tuberculosis*. J Bacteriol 185:6548–6555
35. Ota H, Sakasegawa S, Yasuda Y, Imamura S, Tamura T (2008) A novel nucleoside kinase from *Burkholderia thailandensis*: a member of the phosphofructokinase B-type family of enzymes. FEBS J 275:5865–5872
36. Haase GHW, Brune M, Reinstein J, Pai EF, Pingoud A, Wittinghofer A (1989) Adenylate kinases from thermosensitive *Escherichia coli* strains. J Mol Biol 207:151–162
37. Taniguchi H, Imura M, Okano K, Honda K (2019) Developing a single strain for in vitro salvage synthesis of NAD^+ at high temperatures and its potential for bioconversion. Microb Cell Factories 18:75
38. Fehlau M, Kaspar F, Hellendahl KF, Schollmeyer J, Neubauer P, Wagner A (2020) Modular enzymatic cascade synthesis of nucleotides using a (d)ATP regeneration system. Front Bioeng Biotechnol 8:854
39. Holden HM, Rayment I, Thoden JB (2003) Structure and function of enzymes of the Leloir pathway for galactose metabolism. J Biol Chem 278:43885–43888
40. Kim BH, Gadd GM (2008) Bacterial physiology and metabolism.1st edn. Cambridge University Press, Cambridge
41. Lagunas-Rangel FA (2019) Current role of mammalian sirtuins in DNA repair. DNA Repair 80: 85–92
42. Fukuzumi S, Lee Y, Nam W (2018) Artificial photosynthesis for production of ATP, NAD(P)H, and hydrogen peroxide. ChemPhotoChem 2:121–135
43. Alberts B, Johnson A, Lewis J, Raff M, Roberts K, Walter P (2002) Catalysis and the use of energy by cells. Molecular biology of the cell.4th edn. Garland Science, New York
44. Gonzalo G, Paul CE (2021) Recent trends in synthetic enzymatic cascades promoted by alcohol dehydrogenases. Curr Opin Green Sustain Chem 32:100548

45. Gálvez S, Gadal P (1995) On the function of the NADP-dependent isocitrate dehydrogenase isoenzymes in living organisms. Plant Sci 105:1–14
46. Tommasini-Ghelfi S, Murnan K, Kouri FM, Mahajan AS, May JL, Stegh AH (2019) Cancer-associated mutation and beyond: the emerging biology of isocitrate dehydrogenases in human disease. Sci Adv 5:eaaw4543
47. Begley TP, Kinsland C, Mehl RA, Osterman A, Dorrestein P (2001) The biosynthesis of nicotinamide adenine dinucleotides in bacteria. Vitam Horm 61:103–119
48. Castro-Portuguez R, Sutphin GL (2020) Kynurenine pathway, NAD^+ synthesis, and mitochondrial function: targeting tryptophan metabolism to promote longevity and healthspan. Exp Gerontol 132:110841
49. Preis J, Handler P (1958) Biosynthesis of diphosphopyridine nucleotide: I. Identification of intermediates. J Biol Chem 233:488–492
50. Preis J, Handler P (1958) Biosynthesis of diphosphopyridine nucleotide: II. Enzymatic aspects. J Biol Chem 233:493–500
51. Yaku K, Okabe K, Hikosaka K, Nakagawa T (2018) NAD metabolism in cancer therapeutics. Front Oncol 8:622
52. Walt DR, Rios-Mercadillo VM, Auge J, Whitesides GM (1980) Synthesis of nicotinamide adenine dinucleotide (NAD) from adenosine monophosphate (AMP). J Am Chem Soc 102: 7805–7806
53. Honda K, Hara N, Cheng M, Nakamura A, Mandai K, Okano K, Ohtake H (2016) *In vitro* metabolic engineering for the salvage synthesis of NAD. Metab Eng 35:114–120
54. Grose JH, Joss L, Velick SF, Roth JR (2006) Evidence that feedback inhibition of NAD kinase controls responses to oxidative stress. Proc Natl Acad 103:7601–7606
55. Shianna KV, Marchuk DA, Strand MK (2006) Genomic characterization of POS5, the *Saccharomyces cerevisiae* mitochondrial NADH kinase. Mitochondrion 6:94–101
56. Yang L, Mu X, Nie Y, Xu Y (2021) Improving the production of NAD^+ via multi-strategy metabolic engineering in *Escherichia coli*. Metab Eng 64:122–133
57. Pinson B, Ceschin J, Saint-Marc C, Daignan-Fornier B (2019) Dual control of NAD^+ synthesis by purine metabolites in yeast. elife 8:e43808
58. Zachos I, Döring M, Tafertshofer G, Simon RC, Sieber V (2021) carba Nicotinamide adenine dinucleotide phosphate: robust cofactor for redox biocatalysis. Angew Chem Int Ed 60:14701–14706
59. Hummel W, Gröger H (2014) Strategies for regeneration of nicotinamide coenzymes emphasizing self-sufficient closed-loop recycling systems. J Biotechnol 191:22–31
60. Immanuel S, Sivasubramanian R, Gul R, Dar MA (2020) Recent progress and perspectives on electrochemical regeneration of reduced nicotinamide adenine dinucleotide (NADH). Chem Asian J 15:4256–4270
61. Wu H, Tian C, Song X, Liu C, Yang D, Jiang Z (2020) Methods for the regeneration of nicotinamidecoenzymes. R Soc Chem 15:1773–1789
62. Bornadel A, Hatti-Kaul R, Hollmann F, Kara S (2015) A bi-enzymatic convergent cascade for ε-caprolactone synthesis employing 1,6-hexanediol as a 'double-smart cosubstrate'. ChemCatChem 7:2442–2445
63. Begley TP, Kinsland C, Strauss E (2001) The biosynthesis of coenzyme A in bacteria. Vitam Horm 61:157–171
64. Leonardi R, Zhang Y-M, Rock CO, Jackowski S (2005) Coenzyme A: back in action. Prof Lipid Res 44:125–153
65. Mouterde LMM, Stewart JD (2019) An efficient chemoenzymatic synthesis of coenzyme A and its disulfide. Org Process Res Dev 23:19–30
66. Shimizu S, Komaki R, Tani Y, Yamada H (1983) A high yield method for the preparative synthesis of coenzyme A by combination of chemical and enzymic reactions. FEBS Lett 151: 303–306
67. Chenault HK, Simon ES, Whitesides GM (1988) Cofactor regeneration for enzyme-catalysed synthesis. Biotechnol Genet Eng Rev 6:221–270

68. Crans DC, Whitesides GM (1983) A convenient synthesis of disodium acetyl phosphate for use in *in situ* ATP cofactor regeneration. J Org Chem 48:3130–3132
69. Greunke C, Glöckle A, Antosch J, Gulder TAM (2017) Biocatalytic total synthesis of ikarugamycin. Angew Chem Int Ed 56:4351–4355
70. Hughes AJ, Keatinge-Clay A (2011) Enzymatic extender unit generation for *in vitro* polyketide synthase reactions: structural and functional showcasing of *Streptomyces coelicolor* MatB. Chem Biol 18:165–176
71. Miyazawa T, Hirsch M, Zhang Z, Keatinge-Clay AT (2020) An *in vitro* platform for engineering and harnessing modular polyketide synthases. Nat Commun 11:80
72. Yan Y, Jia P, Bai Y, Fan TP, Zheng X, Cai Y (2019) Production of rosmarinic acid with ATP and CoA double regenerating system. Enzym Microb Technol 131:109392
73. Liscombe DK, Louie GV, Noel JP (2012) Architectures, mechanisms and molecular evolution of natural product methyltransferases. Nat Prod Rep 29:1238–1250
74. Barreiro EJ, Kümmerle AE, Fraga CAM (2011) The methylation effect in medicinal chemistry. Chem Rev 111:5215–5246
75. Zhang J, Klinman JP (2015) High-performance liquid chromatography separation of the (*S,S*)- and (*R,S*)-forms of *S*-adenosyl-L-methionine. Anal Biochem 476:81–83
76. Sofia HJ, Chen G, Hetzler BG, Reyes-Spindola JF, Miller NE (2001) Radical SAM, a novel protein superfamily linking unresolved steps in familiar biosynthetic pathways with radical mechanisms: functional characterization using new analysis and information visualization methods. Nucleic Acids Res 29:1097–1106
77. Mordhorst S, Siegrist J, Müller M, Richter M, Andexer JN (2017) Catalytic alkylation using a cyclic *S*-adenosylmethionine regeneration system. Angew Chem Int Ed 56:4037–4041
78. Liao C, Seebeck FP (2019) *S*-adenosylhomocysteine as a methyl transfer catalyst in biocatalytic methylation reactions. Nat Catal 2:696–701
79. Ospina F, Schülke KH, Soler J, Klein A, Prosenc B, Garcia-Borràs M, Hammer SC (2022) Selective biocatalytic *N*-methylation of unsaturated heterocycles. Angew Chem Int Ed Engl 61: e202213056
80. Wen X, Leisinger F, Leopold V, Seebeck FP (2022) Synthetic reagents for enzyme-catalyzed methylation. Angew Chem Int Ed 61:e202208746

Adv Biochem Eng Biotechnol (2023) 186: 51–76
https://doi.org/10.1007/10_2023_220

Published online: 13 June 2023

Hydrogel-Based Multi-enzymatic System for Biosynthesis

Check for updates

Han Wu and Bo Zheng

Contents

1 Introduction 52
2 Enzyme Immobilization on Hydrogels 54
2.1 Covalent Immobilization of Enzymes on Hydrogels 54
2.2 Non-covalent Immobilization of Enzymes on Hydrogels 59
2.3 Summary of Enzyme Immobilization on Hydrogels 62
3 Applications of Hydrogel-Based Multi-enzymatic System for Biosynthesis 64
3.1 Hydrogel-Based Multi-enzymatic System for Protein Synthesis 64
3.2 Hydrogel-Based Multi-enzymatic Systems for the Synthesis of Non-protein Compounds 66
4 Summary and Outlook 70
References 71

Abstract Biosynthesis involving multi-enzymatic reactions is usually an efficient and economic method to produce plentiful important molecules. To increase the product yield in biosynthesis, the involved enzymes can be immobilized to carriers for enhancing enzyme stability, increasing synthesis efficiency and improving enzyme recyclability. Hydrogels with three-dimensional porous structures and versatile functional groups are promising carriers for enzyme immobilization. Herein, we review the recent advances of the hydrogel-based multi-enzymatic system for biosynthesis. First, we introduce the strategies of enzyme immobilization in hydrogel, including the pros and cons of the strategies. Then we overview the recent applications of the multi-enzymatic system for biosynthesis, including cell-free protein synthesis (CFPS) and non-protein synthesis, especially high value-added molecules. In the last section, we discuss the future perspective of the hydrogel-based multi-enzymatic system for biosynthesis.

H. Wu and B. Zheng (✉)
Institute for Cell Analysis, Shenzhen Bay Laboratory, Shenzhen, Guangdong, China
e-mail: bozheng@szbl.ac.cn

Graphical Abstract

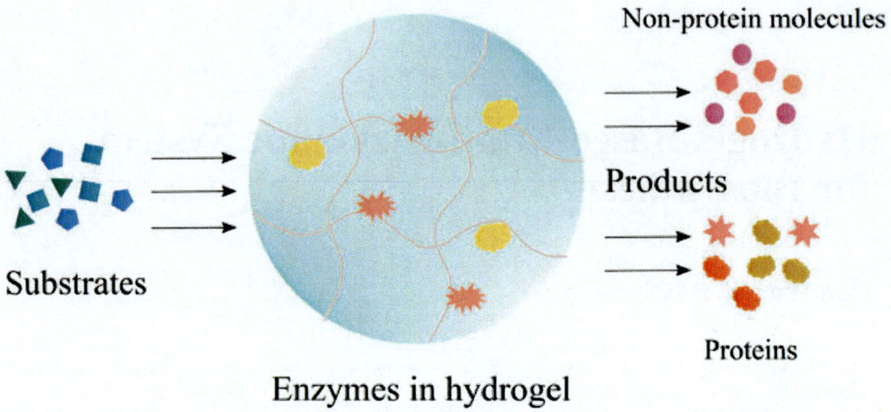

Hydrogels with three-dimensional porous structures and versatile functional groups are promising carriers for enzyme immobilization. The hydrogel-based multi-enzymatic systems are developed for biosynthesis, including cell-free protein synthesis and non-protein synthesis, especially high value-added molecules.

Keywords Bioconjugation, Biosynthesis, Hydrogel

1 Introduction

Biosynthesis is a process involving multi-enzymatic reactions that converts substrates to products. Biosynthesis is widely used to produce pharmaceuticals, biomolecules, industrial compounds, etc. [1–5]. In biosynthesis, successive enzymatic reactions involve a number of enzymes, via a parallel, a tandem, a cascade, or a network configuration and take place inside living cells or in a cell-free format to generate the final products [6, 7]. Compared to biosynthesis in living cells, cell-free biosynthesis mimics the enzymatic reaction pathways of living cells by selecting and assembling the essential enzymes, leading to the simplification of metabolic networks, elimination of the constrains of cell viability and product toxicity, and easy optimization of reaction constitutions [5, 8, 9].

To perform cell-free multi-enzymatic biosynthesis, one-pot reaction which mixes all enzymes in the same batch reactor is conventionally used for its simplicity and easy control of the components. However, the one-pot reaction suffers from selecting enzymes with compatible optimal catalytical conditions, potential crossover and feedback inhibition among different enzymes, and difficulty in enzymes recycling [10–13]. An alternative to resolve the limitations of one-pot synthesis is to immobilize enzymes on solid carriers [10, 11, 14–18]. Enzyme immobilization facilitates easy separation of enzymes from the reaction, leading to recycling the enzymatic system and cost reduction of biosynthesis. Moreover, the immobilization of enzymes could enhance enzyme stability by maintaining their chemical structures, leading to longer enzyme catalytical lifetime and dry storage. Enzymes can be either co-immobilized on the same carrier or immobilized on different carriers, which allows the controlling of reaction conditions of each individual enzyme for longer enzyme lifetime and better product yield.

The benefit of the immobilization of enzymes in biosynthesis is highly dependent on the choice of the solid carrier for immobilization. Silica beads and resin are popular candidates of the solid carrier. However, in the case of multiple enzymes immobilization, the rigid backbone of silica or resin hinders the synergistic interactions of the enzymes. In contrast, the flexible backbone of hydrogel is more advantageous and facilitates the multi-enzymatic system to achieve better overall catalytic power [16, 17, 19].

In addition to the flexible backbone, the solution-like nature of hydrogel provides a desirable environment for enzymes by protecting and stabilizing their structure organizations [16, 17, 20, 21]. Enzymes which are trapped or immobilized in hydrogels exhibit comparable catalytic efficiency as in solutions, and sometimes even higher when organic compounds are presented in the system [22, 23]. The porous three-dimensional structure also makes hydrogel an excellent platform to perform catalytical reactions. The porous structure allows mass transfer between hydrogels and the surrounding solution, making the substrates to interact with enzymes easily. The pore size of hydrogel can be tuned from nm to μm to further control the molecules exchanging behaviors. Moreover, the physically insoluble backbones of the self-contained hydrogel permit easy manipulations during synthesis, such as depletion from solutions for product purification and enzyme reusing.

Besides the above-mentioned common advantageous properties of hydrogels, each type of hydrogel has its own characteristics, which originate from the repeating molecular units of polymer chains [24–26]. The repeating units can be further modified with new functional groups to endow the hydrogel with additional or adjustable properties. Nowadays, hydrogels which are charged, degradable, self-healing, or responsive to pH and temperature stimuli can all be prepared as needed [27–36]. For instance, poly(N-isopropylacrylamide) (PNIPAM) is a thermo-responsive hydrogel which undergoes a reversible lower critical solution temperature (LCST) phase transition [37, 38]. When the enzymes are immobilized in PNIPAM, the pore size of the PNIPAM will change substantially during the LCST phase transition, resulting in transforming enzyme steric structures and further affecting the synthesis process [39, 40].

As a promising platform for biosynthesis, hydrogel-based multi-enzymatic biosynthesis has attracted much attention in recent years [19, 20, 22, 39–44]. In this review, we summarize the recent advances of the hydrogel-based multi-enzymatic system for biosynthesis. First, we introduce the strategies of enzyme immobilization in hydrogel, including the pros and cons of the strategies. Then we overview the recent applications of the multi-enzymatic system for biosynthesis, including cell-free protein synthesis and non-protein synthesis, especially high value-added molecules. In the last section, we discuss the future perspective of the hydrogel-based multi-enzymatic system for biosynthesis.

2 Enzyme Immobilization on Hydrogels

2.1 Covalent Immobilization of Enzymes on Hydrogels

Covalently immobilization of enzymes on hydrogels produces the strongest linkage between the enzymes and the backbone of the hydrogels. Covalent bonds are formed between the reactive groups of the enzyme and the hydrogel backbones. Unlike hydrogels which can be created with a variety of monomers containing different kinds of functional groups, enzymes provide narrow selections of reactive groups for covalent binding. A few amino acids in the enzymes, such as lysine, cysteine, glutamic and aspartic acid, offer reactive groups like amine group, thiol group, and carboxylic acid group for further chemical reactions. Alternatively, in addition to the inherent functional groups of the enzyme, some modified tags linked to the enzymes also contribute to the covalent binding of enzymes via specific interactions between tags and functional groups on the carriers.

2.1.1 Enzymes Without Modification Are Covalently Bound on Hydrogels

There are several common protein bioconjugation techniques to covalently immobilize unmodified enzymes on hydrogels. One such method is to exploit the 1-Ethyl-3-(3-dimethylaminopropyl) carbodiimide (EDC) and N-hydroxy succinimide (NHS) conjugation chemistry, which allows high efficiency coupling of two molecules with corresponding functional groups. Hydrogels with carboxylic acid groups can be activated with carbodiimide and NHS to present amine-reactive esters, which can then spontaneously react with primary amines of enzymes by forming amide bonds (Fig. 1a-i). For instance, Lai et al. [21] (Fig. 1b) immobilized trypsin on poly (N-isopropylacrylamide-acrylic acid) (PNIPAAm-AA) microgels by activating carboxylic acid groups on acrylic acid backbones with EDC and subsequently reacting with amino groups on trypsin to form covalent bonds. It was observed that both the maximum activity (V_{max}) and Michaelis constant (K_m) of the immobilized trypsin increased 1.5-fold compared to the free enzyme. The immobilized trypsin also exhibited better thermal stability and storage stability. With immobilization, the enzyme activity increased from 60 to 91% after 60 min at 45°C and from 55 to 80% after 60 days of storage.

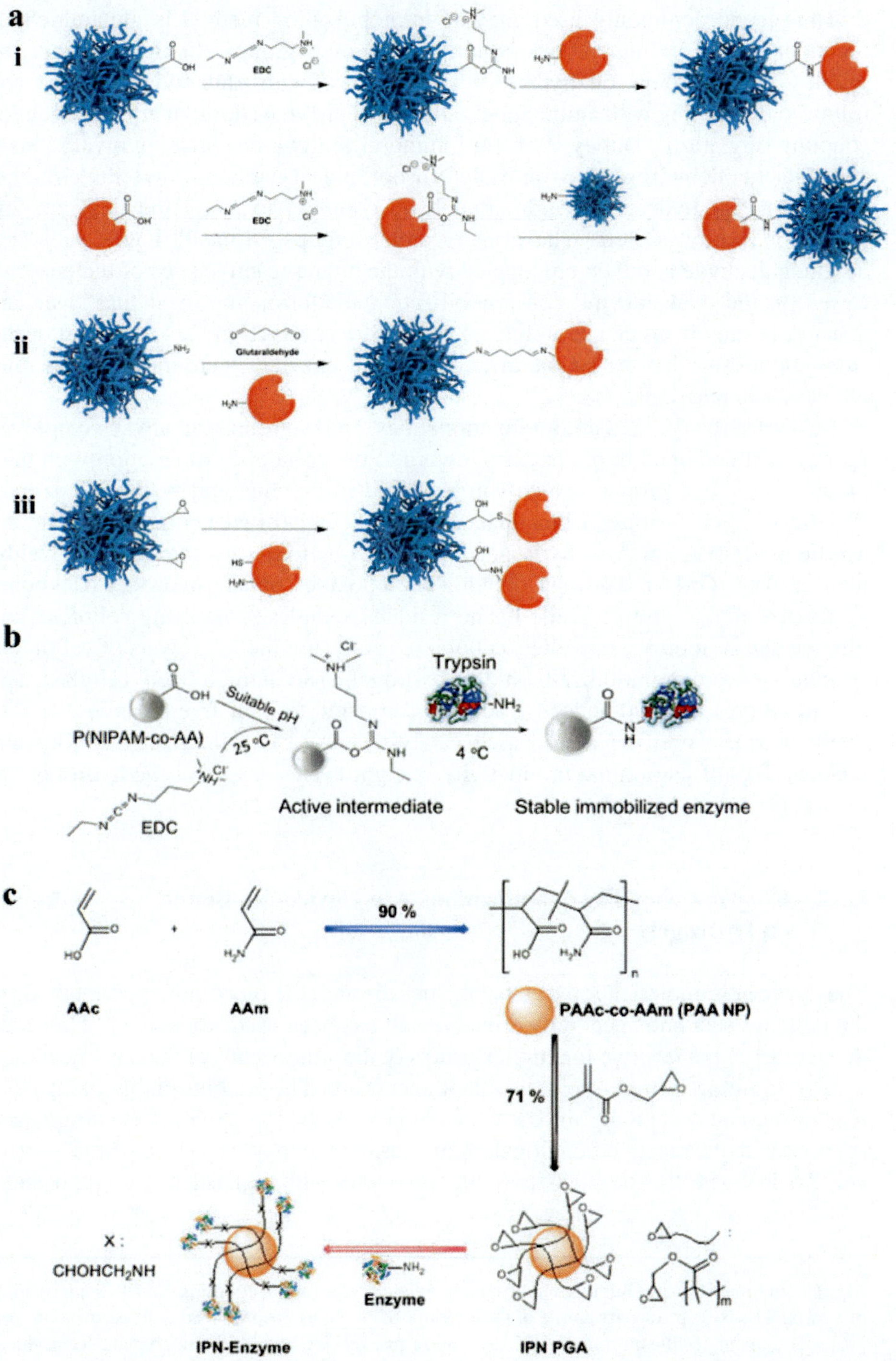

Fig. 1 Enzymes without modification are covalently bound on hydrogels. (**a**) Strategies for covalent enzyme immobilization on hydrogels. (**a-i**) 1-Ethyl-3-(3-dimethylaminopropyl) carbodiimide (EDC) was used to activate carboxylate groups for reacting with amine groups.

The second commonly used enzyme immobilization method is glutaraldehyde functionalization, which allows coupling of amine groups. In this method, the primary amine groups on hydrogels are activated with glutaraldehyde first and followed by reacting with amine functionalities of enzymes through amine–aldehyde coupling (Fig. 1a-ii). Dubey et al. [46] immobilized two enzymes, pyruvate kinase and *L*-lactic dehydrogenase on poly(N-isopropylacrylamide)-poly(ethylenimine) (PNIPAm-PEI) hydrogel via glutaraldehyde coupling to study the dual enzyme co-immobilization system. The primary amine groups on the PEI were activated by glutaraldehyde and then conjugated with the amine-reactive sites of the enzyme. Based on the dual enzyme co-immobilization platform, improved turn over and enhanced catalytic efficiency of the enzymes were observed in the kinetic parameter study, indicating that the spatial arrangements of enzymes were important in catalytic cascade reactions.

Besides carboxyl- and amino functionalities, epoxy groups can also be employed for covalent conjugation of enzymes, owing to their tendency of reaction with thiol groups and amine groups, which can be found in cysteine and lysine in enzymes (Fig. 1a-iii). For example, Limadinata et al. [45] (Fig. 1c) created poly(acrylamide-acrylic acid) (PAAm-AA) hydrogels with epoxy groups by integrating glycidyl methacrylate (GMA). Through opening the epoxides on the hydrogel backbones by the free amine groups in the lysine residues, cellulase catalyzing cellulose into glucose and cellobiose, as well as cellobiase catalyzing the hydrolysis of cellobiose to glucose were immobilized on the hydrogel. The immobilized cellulase and cellobiase preserved 90 and 78% activity compared to their free enzymes, respectively. Furthermore, the as-prepared catalyst showed excellent recyclability and retained 71% of its original productivity in eight cycles usage for the hydrolysis of filter paper.

2.1.2 Enzymes with Tag Modifications Are Covalently Bound on Hydrogels

The covalent immobilization of the enzymes through the bioconjugation methods in the previous section is generally effective and has been used extensively. However, the number of the reactive functional groups of the amino acids of the enzyme varies, usually resulting in multi-point covalent anchoring. The uncontrollable multi-point reactions could potentially impair the enzyme activity, especially if the amino acid residues with the reactive functional groups reside within or near the enzyme's active site. To deal with this problem, coupling processes with high selectivity are desired.

Fig. 1 (continued) (**a-ii**) Glutaraldehyde-mediated aldehyde–amine coupling. (**a-iii**) Ring opening of hydrogel epoxide groups by amine or thiol groups of enzymes. Reproduced with permission from [16]. (**b**) Schematic illustration of EDC activation process for immobilizing trypsin. Reproduced with permission from [21]. (**c**) Procedure of the enzyme immobilization on hydrogel particles through ring opening of epoxide groups. Reproduced with permission from [45]

The highly specific reaction between nickel–nitrilotriacetic acid (Ni-NTA) and histidine tag is widely employed in protein purification, in which the histidine tagged proteins are specifically immobilized on the stationary phase presenting Ni-NTA. Based on the same principle, enzymes modified with histidine tags can also bind onto the hydrogel which contains Ni-NTA in the backbone [19, 47–49]. One example of the histidine tagged (his-tagged) multi-enzymatic system is the Protein synthesis Using Recombinant Elements (PURE) system, which consists of purified proteinaceous factors necessary for transcription and translation [50]. Zhou et al. [19] immobilized the PURE system onto the polyacrylamide (PA) hydrogel for cell-free protein synthesis (Fig. 2a). The PA hydrogel was first polymerized with acrylic acid NHS ester to present NHS groups on the backbone for further conjugation with Nα, Nα-bis(carboxymethyl)-L-lysine hydrate (AB-NTA), and nickel ions. The his-tagged proteinaceous factors of PURE were then immobilized by interacting with Ni-NTA. It was observed that with the continuous supply of feeding buffer containing nutrients and energy, protein synthesis in the hydrogel remained stable for at least 11 days. Gene regulation and gene circuit engineering were further demonstrated on this self-maintaining and self-regulating hydrogel platform.

To decrease the effect of the toxic nickel ions, Lai et al. [51] replaced the Ni-NTA on the PA hydrogel backbones with anti-his-tag aptamers (Fig. 2b). Aptamers are highly biocompatible single strand DNA or RNA molecules with strong binding affinities toward specific molecules [53–55]. The aptamers here were 5′-NH_2 labeled and could react with NHS groups on the polymer scaffold. Subsequently, the his-tagged proteinaceous factors of PURE were immobilized through the aptamer recognition, leading to specific and strong binding. The resulting PA-based multi-enzymatic system afforded continuous and stable protein synthesis for at least 16 days.

Another widely used strategy for protein site-specific conjugation or labeling is sortase-mediated transpeptidation or sortagging, which employs a transpeptidase, Sortase A (SrtA), derived from *Staphylococcus aureus* [52, 56–58]. SrtA can conjugate target molecules with triglycine tag (GGG) and the sorting motif LPXTG by cleaving the amide bond between threonine and glycine residues of LPXTG first and subsequently reacting with nucleophiles of triglycine tags (Fig. 2c-i). Zou et al. [52] successfully employed the sortagging strategy to immobilize enzymes on microgels. In the work, poly(N-vinylcaprolactam)/glycidyl methacrylate (PVCL/GMA) hydrogel was tagged with SrtA recognition peptide sequence LPETG or its nucleophilic counterpart-tag GGG for enzyme conjugation (Fig. 2c). Five different enzymes were successfully immobilized on the hydrogel, which maintained the enzyme activity and supported multiple cycles of usage in biosynthesis.

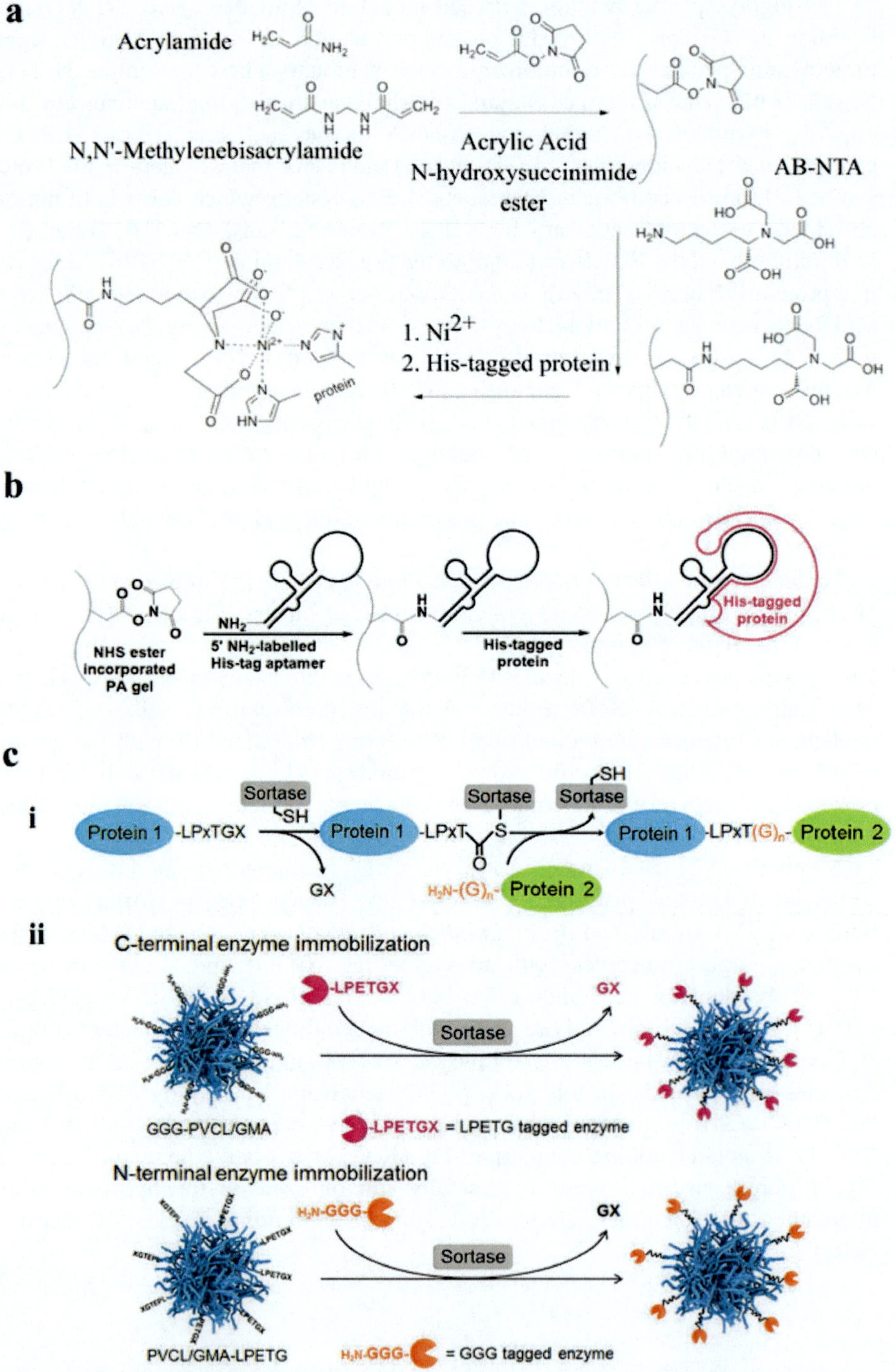

Fig. 2 Enzymes with tag modifications are covalently bound on hydrogels. (**a**) The synthesis of the functionalized PA hydrogel and the immobilization of the his-tagged proteinaceous factors of PURE to the hydrogel scaffold. Reproduced with permission from [19]. (**b**) The schematic of the

2.2 *Non-covalent Immobilization of Enzymes on Hydrogels*

In non-covalent immobilization of enzymes, enzymes are physically encapsulated within the hydrogel or adsorbed onto the polymer backbone of the hydrogel. One simple, fast, and reliable method for non-covalent immobilization is to encapsulate the enzymes during hydrogel polymerization. If the enzymes are premixed with the monomer solution, the enzymes will be encapsulated within the hydrogel networks during the monomer crosslinking process [22, 41, 43, 59, 60]. The enzymes can also be encapsulated in the pre-formed hydrogel by taking advantage of the dehydration and rehydration of mechanically stable hydrogels. Such hydrogels can be dehydrated with heating or freeze-drying to remove the internal water and re-swollen upon contacting with water. Enzymes dissolved in the solution will then be absorbed and encapsulated during the rehydration process [61–63]. In addition, enzymes can also diffuse into hydrogels and be adsorbed onto the hydrogel backbones through physical interactions, such as electrostatic interaction, hydrogen bonding, hydrophobic interactions, etc. [20, 64].

2.2.1 Enzymes Are Encapsulated into Hydrogels During Hydrogel Polymerization

Although encapsulating enzymes in hydrogel during polymerization is easy and feasible, the polymerization strategy should be considered carefully, as most of the conventional polymerization techniques involve harsh reaction conditions, for instance, strong UV illumination, high temperature, presence of surfactants, etc., which could irreversibly denature the enzymes.

Ammonium or potassium persulfate combined with tetramethyl ethylenediamine (TEMED) is one of the most frequently used systems to encapsulate enzymes, as the polymerization process is mild at room temperature [65]. Alginate, which is crosslinked with the addition of Ca^{2+} is another widely used hydrogel for enzyme encapsulation. Yang et al. [66] developed a biosensor by encapsulating enzymes into an alginate–mesoporous silica hydrogel (Fig. 3a). To prepare the hydrogel, glucose oxidase was added into an amino-modified nano-sized mesoporous silica suspension, followed with the addition of the alginate solution. Subsequently, the mixtures were injected into $CaCl_2$ solution, resulting in the alginate hydrogel with glucose oxidase encapsulated inside.

Fig. 2 (continued) fabrication of the anti-his-tag aptamer grafted polyacrylamide hydrogel. Reproduced with permission from [51]. **(c)** Enzyme immobilization on hydrogels using sortagging. **(c-i)** Sortase-mediated protein–protein conjugation using LPETG-tagged and GGG-tagged proteins as substrates. **(c-ii)** Both conjugation options (N- and C-terminal) are feasible for sortase-mediated enzyme immobilization on hydrogels. Reproduced with permission from [52]

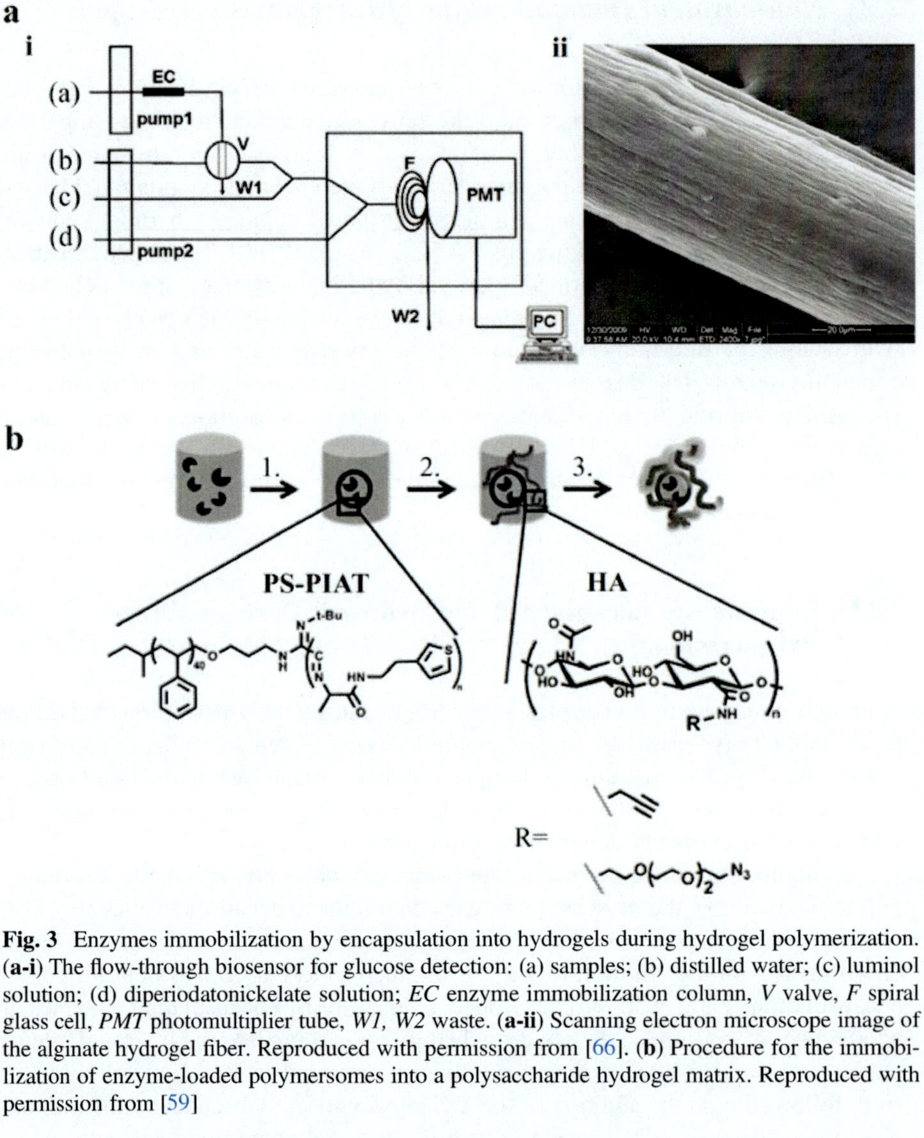

Fig. 3 Enzymes immobilization by encapsulation into hydrogels during hydrogel polymerization. (**a-i**) The flow-through biosensor for glucose detection: (a) samples; (b) distilled water; (c) luminol solution; (d) diperiodatonickelate solution; *EC* enzyme immobilization column, *V* valve, *F* spiral glass cell, *PMT* photomultiplier tube, *W1, W2* waste. (**a-ii**) Scanning electron microscope image of the alginate hydrogel fiber. Reproduced with permission from [66]. (**b**) Procedure for the immobilization of enzyme-loaded polymersomes into a polysaccharide hydrogel matrix. Reproduced with permission from [59]

Click chemistry provides another approach of hydrogel polymerization under mild conditions [67–69], which is useful for enzyme encapsulations. De Hoog et al. [59] constructed a hyaluronic acid-based hydrogel reactor, which encapsulated enzyme containing polymersomes (Fig. 3b). In the work, two enzymes *Candida antarctica* lipase B and glucose oxidase were co-encapsulated in the polymersomes first. Then the polymersomes were added into the azide or acetylene functionalized hyaluronic acid solution. With the addition of the Cu catalyst, the hydrogel was formed with polymersomes trapped inside. Subsequently, a continuous-flow polymersome reactor was built, in which the substrate 2-methoxyphenyl acetate was added to the inlet of the reactor and the product tetraguaiacol was collected at the outlet. The work demonstrated the recycling of the enzymes in repeated reaction cycles without enzyme loss, as the polymersomes and the hydrogel provided double protection against enzyme leaching from the hydrogel.

2.2.2 Enzymes Are Encapsulated into Hydrogels After Polymerization

As the hydrogel polymerization process generally involves harsh conditions that could affect the enzyme stability, enzymes encapsulation is more often performed after the hydrogel polymerization. Some hydrogels can go through the dehydration and rehydration cycles without affecting the mechanical properties, and enzymes can diffuse into the hydrogel during rehydration to be encapsulated. In contrast to the normal diffusing process between the hydrated hydrogel and the surrounding solution, the hydrogel rehydration process allows larger molecules to be encapsulated due to the larger pore size from the dehydration process and achieves a faster absorption kinetics [63].

Whitfield et al. [41] encapsulated the cell extract containing the transcription–translation (TX-TL) system for CFPS in a broad range of hydrogels and studied the effect of hydrogel materials on CFPS. The encapsulation was performed by adding pre-prepared freeze-dried hydrogel into cell extract solutions for rehydration (Fig. 4a-ii). It was observed that many hydrogel materials were compatible with CFPS, including polysaccharide hydrogels, amino acid-based hydrogels, poloxamer gels, and covalently crosslinked hydrogels (Fig. 4b). The yields of the protein synthesis in some hydrogels, such as agarose, agar, and polyacrylamide, were higher

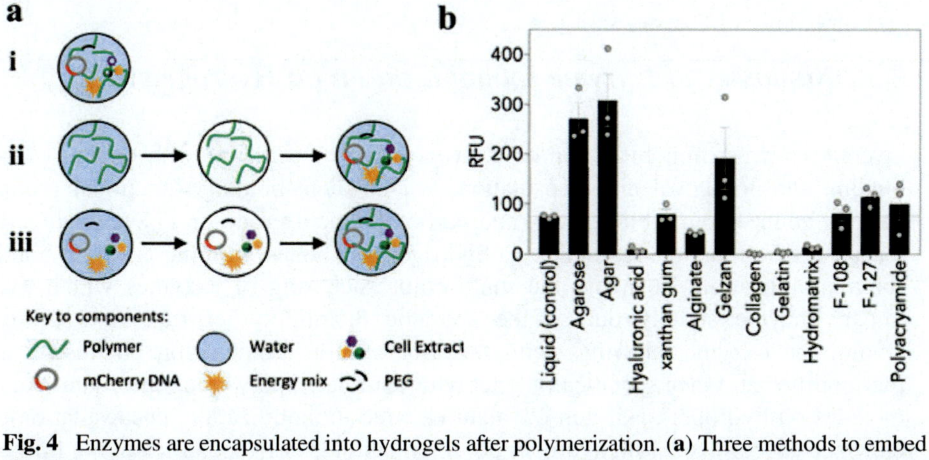

Fig. 4 Enzymes are encapsulated into hydrogels after polymerization. (**a**) Three methods to embed TX-TL system into hydrogels. (**a-i**) TX-TL system was added to solid polymer powders or concentrated liquid hydrogels. (**a-ii**) TX-TL system was added to the dehydrated hydrogel. (**a-iii**) TX-TL system was freeze-dried and then reconstituted with liquid hydrogels. Reproduced with permission from [41]. (**b**) Yields of CFPS in different hydrogel materials loaded with TX-TL system. Reproduced with permission from [41]

than that of the solution-based CFPS because of the crowding effects generated by the hydrogel networks.

To further increase the interactions between enzymes and hydrogel backbones to avoid enzyme leaching, specific physical interactions such as electrostatic interactions, hydrogen bonding, hydrophobic interactions, etc., are introduced in enzyme encapsulation [20, 64]. Dubey et al. [20] prepared a poly(N-isopropylacrylamide)-poly(ethylenimine) (PNIPAm-PEI) hydrogel for acetyl coenzyme A (acetyl CoA) synthetase (Acs) immobilization to synthesize acetyl CoA, which was an essential precursor molecule for the synthesis of metabolites like polyketide-based drugs, lipids, cholesterol, etc. (Fig. 5a). As PEI with the protonated amino groups in the hydrogel was highly charged, enzymes diffused in the hydrogel would be adsorbed onto PEI due to the strong ionic interactions between the cationic and anionic groups on the respective molecules. It was reported that Acs adsorbed in the hydrogel preserved more than 70% of its initial activity after multiple usages and gave a high-rate bioconversion of acetate.

In addition to enzyme adsorption through strong physical interactions, enzymes can also be trapped in stimuli-responsive hydrogels. Hydrogels exhibiting temperature-dependent volume phase transition are often used to encapsulate enzymes. Enzymes can be brought into the polymer networks when the hydrogel is swelling, just like liquid being absorbed into sponges. Once the enzymes are trapped in the swelling hydrogel, the enzymes cannot diffuse out freely. Sigolaeva et al. [64] reported the immobilization of tyrosinase in PNIPAAm-N,N-dimethylaminopropyl methacrylamide (PNIPAAm-DMAPMA) hydrogels by exploiting the swelling and collapsing of the hydrogel upon external stimuli of temperature change (Fig. 5b). A 3.5-fold increase in biosensor sensitivity of phenol assay was achieved on the platform.

2.3 *Summary of Enzyme Immobilization on Hydrogels*

Overall, enzyme immobilization on hydrogels can be achieved with both covalent binding and non-covalent encapsulation. For covalent binding, functional groups such as amine group, thiol group, and carboxylic acid group on enzymes directly react with the functional moieties on hydrogel backbones, forming covalent bonds for immobilization. To avoid the multi-point anchoring of enzymes which may impair enzyme activity due to the sporadic distribution of functional groups among the enzyme, enzymes with tags for specific conjugation chemistry are pre-modified and then specifically react with complementary recognition functional moieties on hydrogels. For non-covalent enzyme immobilization, encapsulation of enzymes concomitant with hydrogel polymerization is easy and efficient but limited to mild polymerization conditions. To avoid catalytical activity loss during hydrogel polymerization with harsh conditions, hydrogels are pre-synthesized and dehydrated. Enzymes are then loaded into hydrogels during the rehydration process for encapsulation. Furthermore, to reduce the enzyme loss after encapsulation in

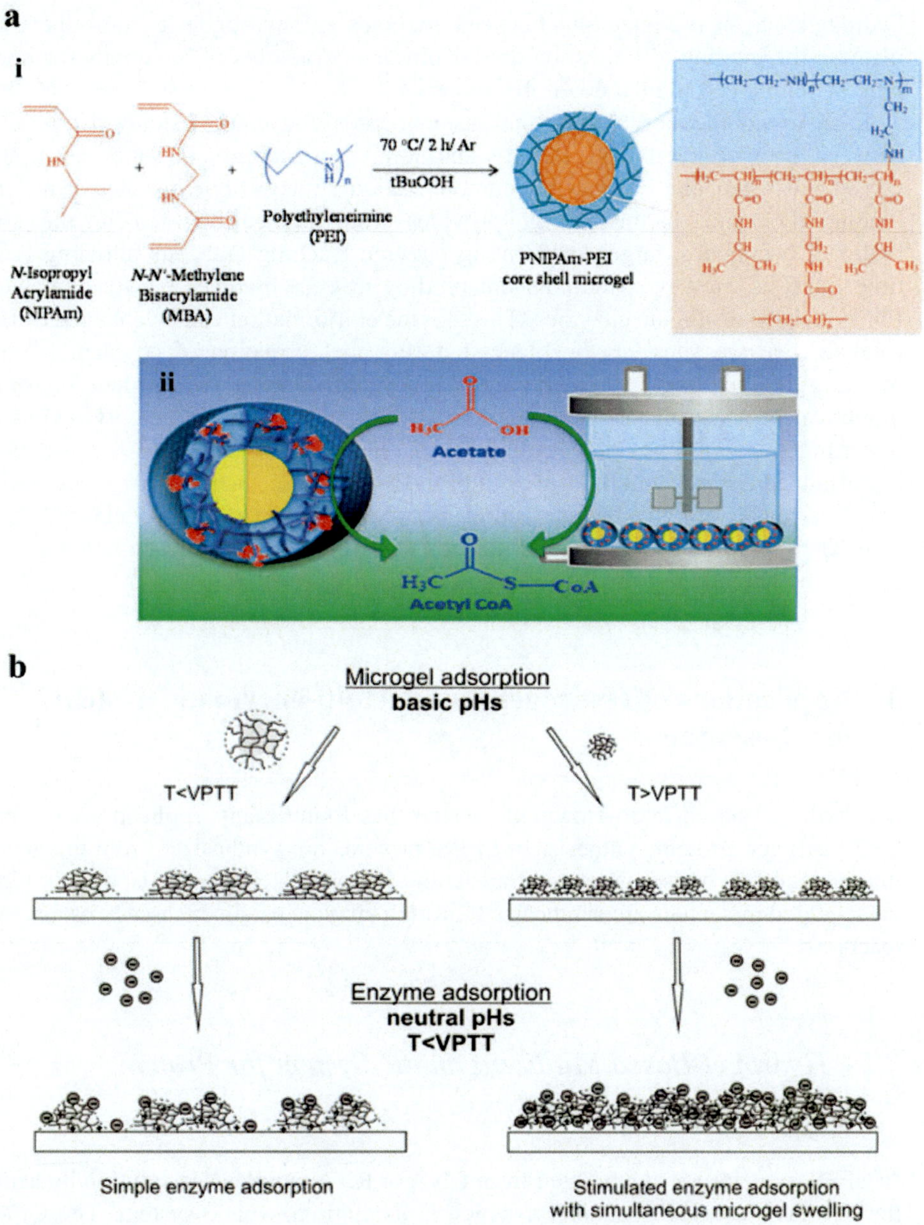

Fig. 5 Enzymes are physically adsorbed or trapped within hydrogels. (**a-i**) Scheme for synthesis of the PNIPAm-PEI hydrogel. (**a-ii**) Schematic illustration of the enzyme immobilized PNIPAm-PEI hydrogel for biocatalysis. Reproduced with permission from [20]. (**b**) Principle of adsorption of enzymes on pH- and temperature-responsive hydrogel. *VPTT* volume phase transition temperature. Reproduced with permission from [64]

hydrogels, physical interactions between enzymes and hydrogels are introduced to increase the binding affinities, or special physical structures of hydrogels for trapping are employed to lock down the enzymes.

Each immobilization strategy has its own pros and cons. Usually a trade-off between enzyme catalytical activity and long-term stability is supposed to be considered when one chooses an immobilization strategy. For instance, covalent binding of enzymes on the hydrogel provides strong linkages between enzymes and hydrogel backbones, largely preventing enzyme leaching and thus allowing long time catalysis. However, the covalent binding process involves the reaction with functional groups on the enzymes. The enzyme conformation can be changed or the catalytic reactive sites can be blocked during the formation of covalent bonds, resulting in lowering the catalytic efficiency. In contrast to covalent binding, non-covalent encapsulation is better suited to retain the enzyme activity owing to fewer functional groups changes of enzymes. On the other hand, the non-covalently encapsulated enzyme could diffuse out of the hydrogel, as the thermal energy of the enzymes could overcome the interactions between enzymes and hydrogel backbones [70]. Therefore, the enzyme immobilization strategy should be chosen based on the specific combination of the enzymes and the hydrogel material.

3 Applications of Hydrogel-Based Multi-enzymatic System for Biosynthesis

The hydrogel-based multi-enzymatic system has found many applications in both CFPS and non-protein synthesis. In CFPS, proteins are synthesized from the genes through cell-free transcription and translation. In non-protein synthesis, usually high value-added molecules are synthesized through enzyme catalytic cascade or tandem reactions.

3.1 Hydrogel-Based Multi-enzymatic System for Protein Synthesis

In CFPS, proteins are synthesized from DNA or RNA templates in solution by using the TX-TL biological machinery, which is a multi-enzymatic system. The CFPS system includes the TX-TL system, the energy generation system, amino acids, nucleotides, and the DNA template [50, 71, 72]. In hydrogel-based multi-enzymatic protein synthesis, some or all of the components of the CFPS system are immobilized in hydrogels.

Benítez-Mateos et al. [43] embedded the CFPS system in hydrogels by casting the alginate–CFPS mixture solution in microchannels to fabricate a micro-compartmentalized hydrogel with subsequently dispensing $CaCl_2$ solution for gelation (Fig. 6a). Proteins with high bioactivity were successfully synthesized on the platform. Due to the porous structure of the hydrogel, the as-produced proteins could diffuse into the bulk solution, making concurrent synthesis and delivery of proteins possible. However, the porous structure also allowed the non-covalently bound CFPS components to diffuse out, resulting in reducing the yield of protein synthesis. To solve this problem, Lim et al. [73] applied a silica coating to the alginate hydrogel particles containing the CFPS system (Fig. 6b). The silica coating acted as a molecular weight cut-off membrane, allowing small molecules to diffuse through while hindering macromolecules such as the TX-TL system from diffusing out. With the silica coating, the productivity of the target protein increased by two-fold and the active lifetime of the CFPS system was elongated by five-fold over the bare alginate particles.

To further increase the protein synthesis yield, the proteinaceous factors of the CFPS system can be immobilized in hydrogels. A typical CFPS is performed in

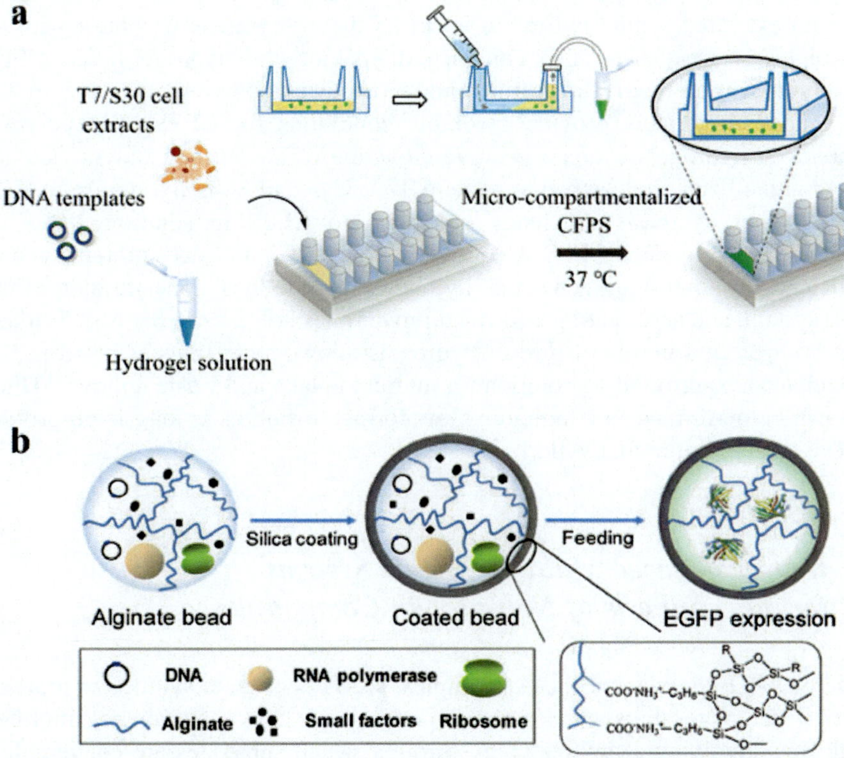

Fig. 6 Hydrogel-based multi-enzymatic system for protein synthesis. Full components of CFPS system were encapsulated in hydrogels. (**a**) CFPS mixtures were encapsulated in alginate hydrogels in microfluidic channels for protein synthesis and concurrent delivery. Reproduced with permission from [43]. (**b**) Silica-coated alginate hydrogel beads encapsulating cell extracts and plasmid DNA for cell-free protein synthesis. *EGFP* enhanced green fluorescent protein. Reproduced with permission from [73]

batch format and lasts only a few hours, due to the accumulation of toxic by-products and the depletion of energy and nutrients [50, 74]. To solve the problem of the CFPS in batch format, CFPS in the format of continuous flow was reported in several studies, which could sustain for much longer time with a continuous supply of energy and nutrients [75–77]. Our group conducted CFPS in the format of continuous flow by immobilizing proteinaceous factors on a PA hydrogel. Zhou et al. [19] immobilized the his-tagged proteinaceous factors of PURE onto the Ni-NTA functionalized PA hydrogel and achieved stable synthesis of mCherry for at least 11 days (Fig. 7a). Lai et al. [51] replaced the Ni-NTA on the PA hydrogel with more biocompatible anti-his-tag aptamers for immobilization of his-tagged proteinaceous factors, and demonstrated the synthesis of mCherry for at least 16 days (Fig. 7b).

The PURE system utilizes T7 RNA polymerase and the means of gene regulatory are limited as a result. Ouyang et al. [78] replaced the PURE system with *E. coli* cell extracts that support more complex gene circuit engineering (Fig. 7c). Here, the components of the cell extracts were covalently immobilized by reacting with the NHS groups on PA hydrogel. Protein products with high bioactivity were synthesized with extended synthesis time of up to 28 days. A genetic oscillator with an active-repressor motif was also constructed, making the hydrogel-based CFPS system promising in biomanufacturing and metabolic engineering.

In general, hydrogels provide favorable conditions for CFPS. The polymer backbones of hydrogels exerts a positive crowding effect on the catalytic activity of the immobilized multi-enzyme system [79–82]. CFPS in hydrogels usually exhibits higher synthesis efficiency compared to CFPS in solutions [41, 43, 73]. Immobilization of the CFPS system in hydrogels also enhances protein synthesis efficiency by improving gene stability, increasing the local concentration of the TX-TL system, and accelerating enzyme turnover rates [19, 51, 78, 83–86]. Furthermore, the open and porous nature of hydrogels allows mass transfer between the hydrogel and the surrounding solution for nutrient uptake and waste removal. Thus, CFPS can be performed in continuous-flow format, resulting in long-lived protein synthesis with high product yield [19, 51, 78].

3.2 Hydrogel-Based Multi-enzymatic Systems for the Synthesis of Non-protein Compounds

Compared to CFPS, which utilizes a complex TX-TL system to synthesize proteins from DNA templates, biosynthesis of non-protein molecules can often be simplified to pathways involving a few enzymes, through which substrates are converted to products [87, 88]. In such cases, purified enzymes with high activity are usually employed. Hydrogels with high water content provide a natural and benign environment and are excellent carriers for enzymes. Enzymes can be immobilized in hydrogels for long-term storage while maintaining their stabilities and catalytical activities [16, 17]. In addition, the ease of separation of synthesized products from

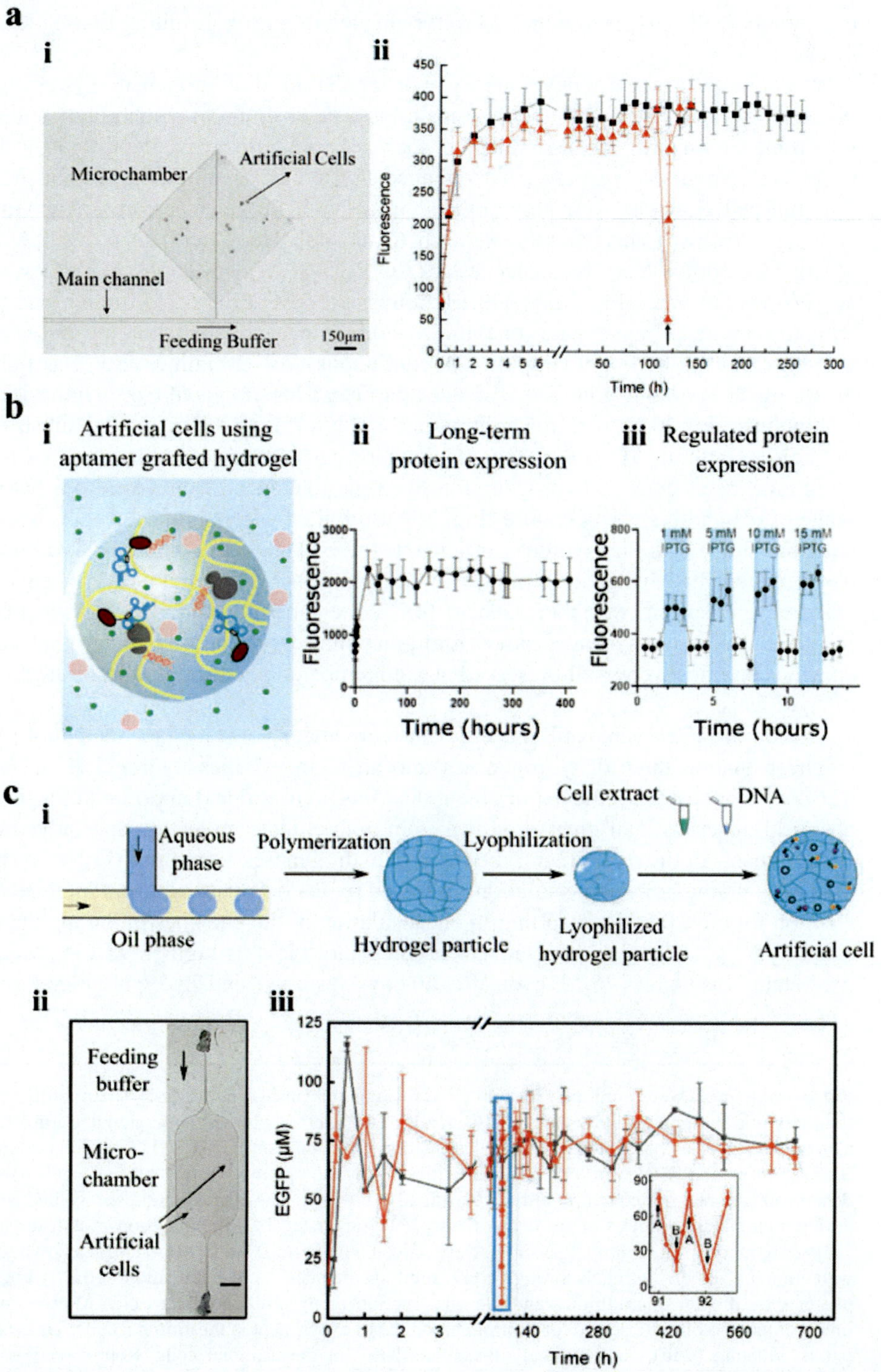

Fig. 7 Hydrogel-based multi-enzymatic system for protein synthesis. Proteinaceous factors of CFPS systems were immobilized on hydrogels. (**a**) His-tagged proteinaceous factors of PURE

the immobilized enzyme system allows for enzyme recycling, resulting in long-term synthesis.

Wu et al. [89] recently developed a hydrogel-based multi-enzymatic system for the economical synthesis of inositol. Inositol is a sugar found in human body and is important in treating mental illnesses such as panic disorder, depression, and obsessive-compulsive disorder. To enhance the synthesis efficiency of inositol, four purified enzymes were immobilized in a PA hydrogel to create an in vitro enzymatic pathway that converted starch to inositol. Here, acrylate polyethylene glycol succinimidyl carboxymethyl ester (AC-PEG-SCM) was copolymerized with acrylamide monomer (Fig. 8). The NHS groups in the AC-PEG-SCM further reacted with the amine groups of the enzyme, resulting in a covalent binding between enzymes and the hydrogel (Fig. 8a). A continuous-flow column reactor was built by filling the column with the hydrogel particles. Inositol could be synthesized continuously for 360 h with the continuous supply of maltodextrin as the substrate through the column (Fig. 8c). The results showed that the immobilization of the multi-enzyme system on hydrogels improved the tolerance of enzymes to a broad range of pH and high temperature (Fig. 8d). Both the hydrogel- and solution-based reactions showed similar synthesis efficiency under pH 8.0–9.0, while the hydrogel-based system had better performance under pH 5.0–7.0. Additionally, when the reaction temperature was increased to 80°C, the inositol yield was significantly increased with the hydrogel-based multi-enzymatic system, indicating that the enzyme activities were well preserved at high temperatures through immobilization on hydrogels.

Xu et al. [22] achieved an efficient conversion of CO_2 to methanol by co-encapsulating three dehydrogenases into an alginate–silica hydrogel (Fig. 9a). The conversion of CO_2, the major greenhouse gas, to useful and important industrial chemical materials is of great significance in both climate protection and industrial development. In the experiment, the three dehydrogenases were premixed with the alginate–tetramethoxysilane solution, followed by the dropwise addition into $CaCl_2$ solution for crosslinking, resulting in encapsulation of the enzymes in the alginate–silica hydrogel. The system achieved the methanol yield as high as 98.1%, which was retained as high as 76.2% even after 60 days of storage and 78.5% after ten times

Fig. 7 (continued) were immobilized with Ni-NTA moieties on hydrogels. **(a-i)** The microfluidic setup for the continuous protein synthesis. **(a-ii)** The average fluorescence signal during the continuous synthesis of mCherry. Reproduced with permission from [19]. **(b)** His-tagged proteinaceous factors of PURE were immobilized with anti-his-tag aptamer grafted on hydrogels. **(b-i)** Scheme of aptamer grafted hydrogel-based artificial cells. **(b-ii)** The average fluorescence intensity during continuous synthesis of mCherry. **(b-iii)** IPTG-regulated EGFP expression in anti-his-tag aptamer functionalized hydrogel particles. Reproduced with permission from [51]. **(c)** Cell extracts were immobilized on hydrogels through NHS coupling chemistry. **(c-i)** Fabrication of the hydrogel particles by droplet microfluidics and the construction of the artificial cells. **(c-ii)** Microscopic image of the microfluidic device for continuous synthesis experiment in the artificial cells. The scale bar is 500 μm. **(c-iii)** Continuous synthesis of EGFP in the artificial cells. Reproduced with permission from [78]

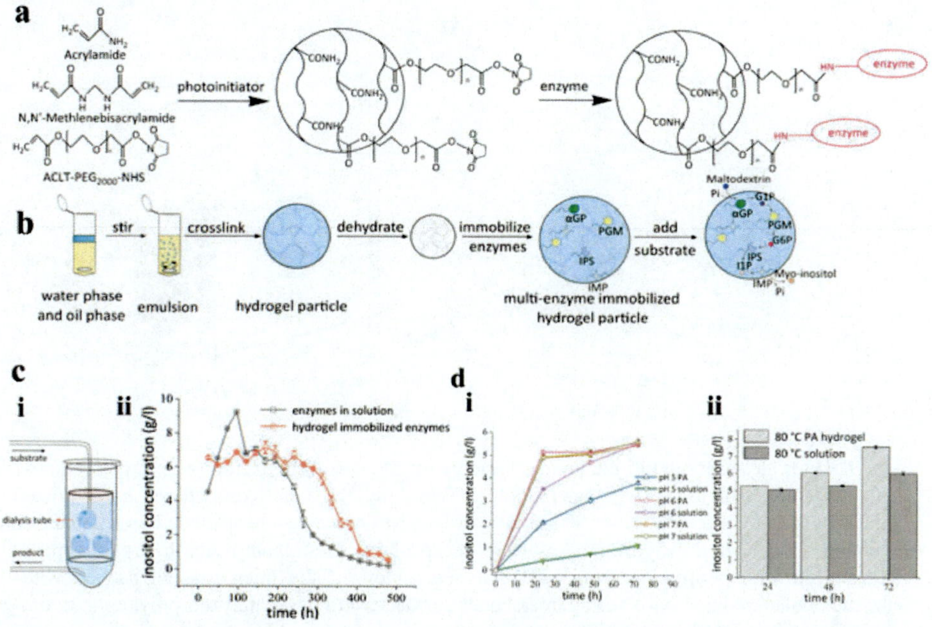

Fig. 8 Continuous biosynthesis of inositol by hydrogel-based multi-enzymatic system. (**a**) Formation principle of the polyacrylamide hydrogel particles. Reproduced with permission from [89]. (**b**) The procedure of enzyme immobilization in hydrogel particles. Reproduced with permission from [89]. (**c-i**) Scheme of the dialysis tube-based device for continuous-flow synthesis. (**c-ii**) The performance of the multi-enzymatic system either immobilized in hydrogel particles or in solution phase for the inositol synthesis for 480 h at 70°C. Reproduced with permission from [89]. (**d**) The inositol yields of the multi-enzymatic system either immobilized in hydrogel particles or in solution phase (**d-i**) at pH 5.0–7.0 and (**d-ii**) at 80°C. Reproduced with permission from [89]

of recycling. It was believed that the appropriate immobilizing microenvironment created by the alginate–silica hydrogel, such as the aqueous medium, moderate rigidity and flexibility of the polymer backbone, ideal diffusion characteristics, as well as optimized cage confinement effect, contributed to the improved long-term enzyme catalytical performance.

In addition, Srinivasan et al. [44] utilized the Ni-NTA functionalized agarose beads to immobilize a his-tagged enzyme, the P450 hydroxylase from the pikromycin polyketide pathway, to catalyze YC-17 to methymycin and neomethymycin, which are 12-membered macrolide antibiotics (Fig. 9b). The results showed that the immobilized enzymes were more stable than those in solution, with an observed half-life increase from ~3 to ~9 h. By assembling the enzyme immobilized agarose beads into a microchannel, the rapid hydroxylation of the macrolide YC-17 to methymycin and neomethymycin with a conversion rate of over 90% was obtained.

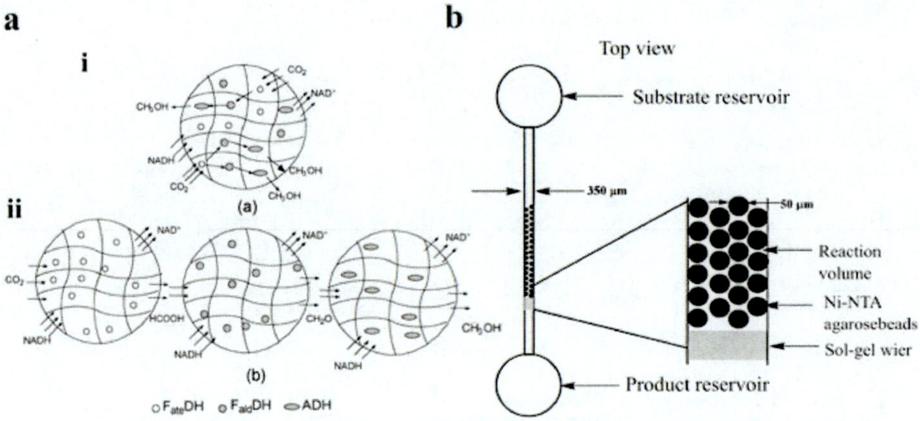

Fig. 9 Hydrogel-based multi-enzymatic systems for the synthesis of non-protein compounds. (**a**) Efficient conversion of CO_2 to methanol catalyzed by three dehydrogenases encapsulated in alginate–silica hybrid gels. (**a-i**) Three dehydrogenases were co-encapsulated in alginate–silica hydrogel. (**a-ii**) Three dehydrogenases were encapsulated in alginate–silica hydrogel separately. Reproduced with permission from [22]. (**b**) Setup view of the microchannel packed with the enzymes functionalized Ni-NTA agarose beads for bacterial P450 catalyzed polyketide hydroxylation. Reproduced with permission from [44]

4 Summary and Outlook

Immobilization of enzymes on hydrogel provides the benefits of improved enzyme stability, enzymes recycling, simplified downstream processing, and continuous-flow format of production. We reviewed the two approaches for enzyme immobilization in hydrogel, i.e. covalent and non-covalent binding. For covalent binding, enzymes can be directly immobilized in hydrogels by exploiting functional groups on the amino acids of the enzymes. Enzymes can also be modified with specific tags, usually at the terminal amino acids, for specific binding with the corresponding functional entities on hydrogel. For non-covalent binding, enzymes can be added to hydrogel monomer solutions and encapsulated into the hydrogel upon hydrogel polymerization. If the hydrogel polymerization is under harsh conditions, enzymes can be encapsulated by passive diffusion into pre-formed hydrogel, or by solution uptake during the rehydration of the dried hydrogel. Physical interactions such as electrostatic interactions, hydrophobic interactions, hydrogen bonding, and physical trapping are employed to minimize or slow down the enzyme leaching from the hydrogel.

We reviewed the applications of the hydrogel-based multi-enzymatic systems for biosynthesis, including the cell-free protein synthesis and non-protein synthesis for high value-added molecules. Overall, hydrogels are promising carriers for multi-enzymatic biosynthesis due to the flexible backbone, biocompatibility, and tunable pore structure. The diversity of hydrogel materials, such as stimuli-responsive hydrogels, endows unique and beneficial properties to the multi-enzymatic systems for biosynthesis [16, 39, 40].

There are a few key issues when one immobilizes enzymes on hydrogel for biosynthesis. Firstly, the catalytic activity of the enzymes could be impaired upon enzyme immobilization, especially in the case of covalent immobilization. Non-covalent immobilization tends to be a more gentle approach, but suffers from enzyme leaching.

Secondly, immobilized enzymes in general pose limitations in mass transfer compared to the free enzymes in solution phase. As a result, the effective specific activity is lowered [90]. The situation will be more severe if the product is in the solid state, which could block the pores of the hydrogel.

Thirdly, similar to other immobilized enzymatic system, the hydrogel-based multi-enzymatic systems for biosynthesis are mostly limited to research laboratories [90, 91]. The critical factor of commercializing the hydrogel-based multi-enzymatic system is the associated cost reduction in the overall production economics. To justify the industrial application of the hydrogel-based multi-enzymatic system for biosynthesis, the cost of the enzyme immobilization must be significantly lower compared to the cost savings by switching from the solution phase to the immobilized multi-enzymatic system.

We are optimistic that the above issues and challenges will be overcome in near future with the advancement of polymer engineering, nanotechnology, and enzyme evolution. The hydrogel-based multi-enzymatic system has great potential not only in biosynthesis, but also in other fields such as biosensor and drug delivery.

Acknowledgments The authors thank the financial support from Shenzhen Bay Laboratory (No. 21280031 and S211101001-1).

References

1. Tibrewal N, Tang Y (2014) Biocatalysts for natural product biosynthesis. Annu Rev Chem Biomol Eng 5:347–366. https://doi.org/10.1146/annurev-chembioeng-060713-040008
2. O'Neill EC, Kelly S (2017) Engineering biosynthesis of high-value compounds in photosynthetic organisms. Crit Rev Biotechnol 37:779–802. https://doi.org/10.1080/07388551.2016.1237467
3. Srinivasulu C, Ramgopal M, Ramanjaneyulu G et al (2018) Syringic acid (SA) – a review of its occurrence, biosynthesis, pharmacological and industrial importance. Biomed Pharmacother 108:547–557. https://doi.org/10.1016/J.BIOPHA.2018.09.069
4. Siddiqi KS, Husen A, Rao RAK (2018) A review on biosynthesis of silver nanoparticles and their biocidal properties. J Nanobiotechnol 16:1–28. https://doi.org/10.1186/s12951-018-0334-5
5. Li J, Zhang L, Liu W (2018) Cell-free synthetic biology for in vitro biosynthesis of pharmaceutical natural products. Synth Syst Biotechnol 3:83–89. https://doi.org/10.1016/j.synbio.2018.02.002
6. Walsh CT, Moore BS (2019) Enzymatic cascade reactions in biosynthesis. Angew Chem Int Ed 58:6846–6879. https://doi.org/10.1002/ANIE.201807844

7. Ricca E, Brucher B, Schrittwieser JH (2011) Multi-enzymatic cascade reactions: overview and perspectives. Adv Synth Catal 353:2239–2262. https://doi.org/10.1002/ADSC.201100256
8. Tang S, Liao D, Li X et al (2021) Cell-free biosynthesis system: methodology and perspective of in vitro efficient platform for pyruvate biosynthesis and transformation. ACS Synth Biol 10: 2417–2433. https://doi.org/10.1021/acssynbio.1c00252
9. Zhuang L, Huang S, Liu WQ et al (2020) Total in vitro biosynthesis of the nonribosomal macrolactone peptide valinomycin. Metab Eng 60:37–44. https://doi.org/10.1016/j.ymben.2020.03.009
10. Mateo C, Palomo JM, Fernandez-Lorente G et al (2007) Improvement of enzyme activity, stability and selectivity via immobilization techniques. Enzym Microb Technol 40:1451–1463. https://doi.org/10.1016/j.enzmictec.2007.01.018
11. Datta S, Christena LR, Rajaram YRS (2013) Enzyme immobilization: an overview on techniques and support materials. Biotech 3:1–9. https://doi.org/10.1007/s13205-012-0071-7
12. Ardao I, Hwang ET, Zeng AP (2013) In vitro multienzymatic reaction systems for biosynthesis. In: Zeng A-P (ed) Advances in biochemical engineering/biotechnology. Springer, Berlin, pp 153–184
13. Morgado G, Gerngross D, Roberts TM, Panke S (2018) Synthetic biology for cell-free biosynthesis: fundamentals of designing novel in vitro multi-enzyme reaction networks. In: Zhao H, Zeng A-P (eds) Advances in biochemical engineering/biotechnology. Springer, Cham, pp 117–146
14. Chapman J, Ismail AE, Dinu CZ (2018) Industrial applications of enzymes: recent advances, techniques, and outlooks. Catalysts 8:238. https://doi.org/10.3390/catal8060238
15. Sheldon RA (2007) Enzyme immobilization: the quest for optimum performance. Adv Synth Catal 349:1289–1307. https://doi.org/10.1002/adsc.200700082
16. Nöth M, Gau E, Jung F et al (2020) Biocatalytic microgels (μ-gel: zymes): synthesis, concepts, and emerging applications. Green Chem 22:8183–8209. https://doi.org/10.1039/d0gc03229h
17. Meyer J, Meyer LE, Kara S (2022) Enzyme immobilization in hydrogels: a perfect liaison for efficient and sustainable biocatalysis. Eng Life Sci 22:165–177. https://doi.org/10.1002/elsc.202100087
18. Liese A, Hilterhaus L (2013) Evaluation of immobilized enzymes for industrial applications. Chem Soc Rev 42:6236–6249. https://doi.org/10.1039/c3cs35511j
19. Zhou X, Wu H, Cui M et al (2018) Long-lived protein expression in hydrogel particles: towards artificial cells. Chem Sci 9:4275–4279. https://doi.org/10.1039/c8sc00383a
20. Dubey NC, Tripathi BP, Müller M et al (2015) Enhanced activity of acetyl CoA synthetase adsorbed on smart microgel: an implication for precursor biosynthesis. ACS Appl Mater Interfaces 7:1500–1507. https://doi.org/10.1021/am5063376
21. Lai E, Wang Y, Wei Y et al (2016) Covalent immobilization of trypsin onto thermo-sensitive poly(N-isopropylacrylamide-co-acrylic acid) microspheres with high activity and stability. J Appl Polym Sci 133. https://doi.org/10.1002/app.43343
22. Xu SW, Lu Y, Li J et al (2006) Efficient conversion of CO2 to methanol catalyzed by three dehydrogenases co-encapsulated in an alginate-silica (ALG-SiO2) hybrid gel. Ind Eng Chem Res 45:4567–4573. https://doi.org/10.1021/ie051407l
23. Gawlitza K, Georgieva R, Tavraz N et al (2013) Immobilization of water-soluble HRP within poly-N-isopropylacrylamide microgel particles for use in organic media. Langmuir 29:16002–16009. https://doi.org/10.1021/la403598s
24. Ahmed EM (2015) Hydrogel: preparation, characterization, and applications: a review. J Adv Res 6:105–121. https://doi.org/10.1016/J.JARE.2013.07.006
25. Kopeček J (2007) Hydrogel biomaterials: a smart future? Biomaterials 28:5185–5192. https://doi.org/10.1016/J.BIOMATERIALS.2007.07.044
26. Fu J, In Het Panhuis M (2019) Hydrogel properties and applications. J Mater Chem B 7:1523–1525. https://doi.org/10.1039/C9TB90023C

27. Kühn S, Freyse J, Atallah P et al (2021) Tuning the network charge of biohybrid hydrogel matrices to modulate the release of SDF-1. Biol Chem 402:1453–1464. https://doi.org/10.1515/hsz-2021-0175
28. Zhang J, Zhu Y, Song J et al (2018) Novel balanced charged alginate/PEI polyelectrolyte hydrogel that resists foreign-body reaction. ACS Appl Mater Interfaces 10:6879–6886. https://doi.org/10.1021/acsami.7b17670
29. Zhao W, Jin X, Cong Y et al (2013) Degradable natural polymer hydrogels for articular cartilage tissue engineering. J Chem Technol Biotechnol 88:327–339. https://doi.org/10.1002/jctb.3970
30. Chen YM, Yang JJ, Gong JP (2009) Adhesion, spreading, and proliferation of endothelial cells on charged hydrogels. J Adhes 85:839–868. https://doi.org/10.1080/00218460903291486
31. Ulijn RV, Bibi N, Jayawarna V et al (2007) Bioresponsive hydrogels. Mater Today 10:40–48. https://doi.org/10.1016/S1369-7021(07)70049-4
32. Kopeček J, Yang J (2007) Hydrogels as smart biomaterials. Polym Int 56:1078–1098. https://doi.org/10.1002/pi.2253
33. Tokarev I, Minko S (2009) Stimuli-responsive hydrogel thin films. Soft Matter 5:511–524. https://doi.org/10.1039/b813827c
34. Koetting MC, Peters JT, Steichen SD, Peppas NA (2015) Stimulus-responsive hydrogels: theory, modern advances, and applications. Mater Sci Eng R Rep 93:1–49. https://doi.org/10.1016/j.mser.2015.04.001
35. Chatterjee S, Hui PCL (2021) Review of applications and future prospects of stimuli-responsive hydrogel based on thermo-responsive biopolymers in drug delivery systems. Polymers (Basel) 13:2086. https://doi.org/10.3390/polym13132086
36. Jiang Y, Wang Y, Li Q et al (2019) Natural polymer-based stimuli-responsive hydrogels. Curr Med Chem 27:2631–2657. https://doi.org/10.2174/0929867326666191122144916
37. Liu R, Fraylich M, Saunders BR (2009) Thermoresponsive copolymers: from fundamental studies to applications. Colloid Polym Sci 287:627–643. https://doi.org/10.1007/S00396-009-2028-X/METRICS
38. Tang L, Wang L, Yang X et al (2021) Poly(N-isopropylacrylamide)-based smart hydrogels: design, properties and applications. Prog Mater Sci 115:100702. https://doi.org/10.1016/J.PMATSCI.2020.100702
39. Wiese S, Spiess AC, Richtering W (2013) Microgel-stabilized smart emulsions for biocatalysis. Angew Chem Int Ed 52:576–579. https://doi.org/10.1002/anie.201206931
40. Dubey NC, Tripathi BP, Stamm M, Ionov L (2014) Smart core-shell microgel support for acetyl coenzyme A synthetase: a step toward efficient synthesis of polyketide-based drugs. Biomacromolecules 15:2776–2783. https://doi.org/10.1021/bm5006382
41. Whitfield CJ, Banks AM, Dura G et al (2020) Cell-free protein synthesis in hydrogel materials. Chem Commun 56:7108–7111. https://doi.org/10.1039/d0cc02582h
42. Heida T, Köhler T, Kaufmann A et al (2020) Cell-free protein synthesis in bifunctional hyaluronan microgels: a strategy for in situ immobilization and purification of his-tagged proteins. ChemSystemsChem 2:e1900058. https://doi.org/10.1002/syst.201900058
43. Benítez-Mateos AI, Zeballos N, Comino N et al (2020) Microcompartmentalized cell-free protein synthesis in hydrogel μ-channels. ACS Synth Biol 9:2971–2978. https://doi.org/10.1021/acssynbio.0c00462
44. Srinivasan A, Bach H, Sherman DH, Dordick JS (2004) Bacterial P450-catalyzed polyketide hydroxylation on a microfluidic platform. Biotechnol Bioeng 88:528–535. https://doi.org/10.1002/bit.20285
45. Limadinata PA, Li A, Li Z (2015) Temperature-responsive nanobiocatalysts with an upper critical solution temperature for high performance biotransformation and easy catalyst recycling: efficient hydrolysis of cellulose to glucose. Green Chem 17:1194–1203. https://doi.org/10.1039/C4GC01742K
46. Dubey NC, Tripathi BP, Müller M et al (2016) Bienzymatic sequential reaction on microgel particles and their cofactor dependent applications. Biomacromolecules 17:1610–1620. https://doi.org/10.1021/acs.biomac.5b01745

47. Byun JY, Lee KH, Lee KY et al (2013) In-gel expression and in situ immobilization of proteins for generation of three dimensional protein arrays in a hydrogel matrix. Lab Chip 13:886–891. https://doi.org/10.1039/C2LC41137G
48. Köhler T, Heida T, Hoefgen S et al (2020) Cell-free protein synthesis and in situ immobilization of deGFP-MatB in polymer microgels for malonate-to-malonyl CoA conversion. RSC Adv 10: 40588–40596. https://doi.org/10.1039/D0RA06702D
49. Lee KH, Lee KY, Byun JY et al (2012) On-bead expression of recombinant proteins in an agarose gel matrix coated on a glass slide. Lab Chip 12:1605–1610. https://doi.org/10.1039/C2LC21239K
50. Shimizu Y, Inoue A, Tomari Y et al (2001) Cell-free translation reconstituted with purified components. Nat Biotechnol 19:751–755. https://doi.org/10.1038/90802
51. Lai SN, Zhou X, Ouyang X et al (2020) Artificial cells capable of long-lived protein synthesis by using aptamer grafted polymer hydrogel. ACS Synth Biol 9:76–83. https://doi.org/10.1021/acssynbio.9b00338
52. Zou Z, Gau E, El-Awaad I et al (2019) Selective functionalization of microgels with enzymes by sortagging. Bioconjug Chem 30:2859–2869. https://doi.org/10.1021/acs.bioconjchem.9b00568
53. Keefe AD, Pai S, Ellington A (2010) Aptamers as therapeutics. Nat Rev Drug Discov 97(9): 537–550. https://doi.org/10.1038/nrd3141
54. Kökpinar Ö, Walter JG, Shoham Y et al (2011) Aptamer-based downstream processing of his-tagged proteins utilizing magnetic beads. Biotechnol Bioeng 108:2371–2379. https://doi.org/10.1002/BIT.23191
55. Tsuji S, Tanaka T, Hirabayashi N et al (2009) RNA aptamer binding to polyhistidine-tag. Biochem Biophys Res Commun 386:227–231. https://doi.org/10.1016/J.BBRC.2009.06.014
56. Popp MW, Antos JM, Grotenbreg GM et al (2007) Sortagging: a versatile method for protein labeling. Nat Chem Biol 311(3):707–708. https://doi.org/10.1038/nchembio.2007.31
57. Nöth M, Zou Z, El-Awaad I et al (2021) A peptide-based coating toolbox to enable click chemistry on polymers, metals, and silicon through sortagging. Biotechnol Bioeng 118:1520–1530. https://doi.org/10.1002/BIT.27666
58. Ritzefeld M (2014) Sortagging: a robust and efficient chemoenzymatic ligation strategy. Chem – A Eur J 20:8516–8529. https://doi.org/10.1002/CHEM.201402072
59. De Hoog HPM, Arends IWCE, Rowan AE et al (2010) A hydrogel-based enzyme-loaded polymersome reactor. Nanoscale 2:709–716. https://doi.org/10.1039/b9nr00325h
60. Baltussen MG, Van De Wiel J, Fernández Regueiro CL et al (2022) A Bayesian approach to extracting kinetic information from artificial enzymatic networks. Anal Chem. https://doi.org/10.1021/acs.analchem.2c00659
61. Liu H, Nakagawa K, Kato DI et al (2011) Enzyme encapsulation in freeze-dried bionanocomposites prepared from chitosan and xanthan gum blend. Mater Chem Phys 129: 488–494. https://doi.org/10.1016/J.MATCHEMPHYS.2011.04.043
62. Wang Y, Shah V, Lu A et al (2021) Counting of enzymatically amplified affinity reactions in hydrogel particle-templated drops. Lab Chip 21:3438–3448. https://doi.org/10.1039/D1LC00344E
63. Li Y, Guo D, Zheng B (2012) Rehydratable gel for rapid loading of nanoliter solution and its application in protein crystallization. RSC Adv 2:4857–4863. https://doi.org/10.1039/C2RA20511D/
64. Sigolaeva LV, Gladyr SY, Gelissen APH et al (2014) Dual-stimuli-sensitive microgels as a tool for stimulated spongelike adsorption of biomaterials for biosensor applications. Biomacromolecules 15:3735–3745. https://doi.org/10.1021/bm5010349
65. Bahar T, Tuncel A (2002) Immobilization of invertase onto crosslinked poly (p-chloromethylstyrene) beads. J Appl Polym Sci 83:1268–1279. https://doi.org/10.1002/APP.2294
66. Yang C, Zhang Z, Shi Z et al (2010) Development of a novel enzyme reactor and application as a chemiluminescence flow-through biosensor. Anal Bioanal Chem 3977(397):2997–3003. https://doi.org/10.1007/S00216-010-3805-6

67. Li X, Xiong Y (2022) Application of "click" chemistry in biomedical hydrogels. ACS Omega 7: 36918–36928. https://doi.org/10.1021/acsomega.2c03931
68. Malkoch M, Vestberg R, Gupta N et al (2006) Synthesis of well-defined hydrogel networks using click chemistry. Chem Commun:2774–2776. https://doi.org/10.1039/b603438a
69. Yigit S, Sanyal R, Sanyal A (2011) Fabrication and functionalization of hydrogels through "click" chemistry. Chem – An Asian J 6:2648–2659. https://doi.org/10.1002/asia.201100440
70. Pardee K, Green AA, Ferrante T et al (2014) Paper-based synthetic gene networks. Cell 159: 940–954. https://doi.org/10.1016/j.cell.2014.10.004
71. Carlson ED, Gan R, Hodgman CE, Jewett MC (2012) Cell-free protein synthesis: applications come of age. Biotechnol Adv 30:1185–1194. https://doi.org/10.1016/J.BIOTECHADV.2011.09.016
72. Lu Y (2017) Cell-free synthetic biology: engineering in an open world. Synth Syst Biotechnol 2:23–27. https://doi.org/10.1016/J.SYNBIO.2017.02.003
73. Lim SY, Kim KO, Kim DM, Park CB (2009) Silica-coated alginate beads for in vitro protein synthesis via transcription/translation machinery encapsulation. J Biotechnol 143:183–189. https://doi.org/10.1016/J.JBIOTEC.2009.07.006
74. Zemella A, Thoring L, Hoffmeister C, Kubick S (2015) Cell-free protein synthesis: pros and cons of prokaryotic and eukaryotic systems. Chembiochem 16:2420–2431. https://doi.org/10.1002/CBIC.201500340
75. Noireaux V, Libchaber A (2004) A vesicle bioreactor as a step toward an artificial cell assembly. Proc Natl Acad Sci U S A 101:17669–17674. https://doi.org/10.1073/pnas.0408236101
76. Niederholtmeyer H, Stepanoèa È, Maerkl SJ (2013) Implementation of cell-free biological networks at steady state. Proc Natl Acad Sci U S A 110:15985–15990. https://doi.org/10.1073/pnas.1311166110
77. Spirin AS, Baranov VI, Ryabova LA et al (1988) A continuous cell-free translation system capable of producing polypeptides in high yield. Science 242:1162–1164. https://doi.org/10.1126/science.3055301
78. Ouyang X, Zhou X, Lai SN et al (2021) Immobilization of proteins of cell extract to hydrogel networks enhances the longevity of cell-free protein synthesis and supports gene networks. ACS Synth Biol 10:749–755. https://doi.org/10.1021/acssynbio.0c00541
79. Morelli MJ, Allen RJ, Rein Ten Wolde P (2011) Effects of macromolecular crowding on genetic networks. Biophys J 101:2882–2891. https://doi.org/10.1016/j.bpj.2011.10.053
80. Gnutt D, Ebbinghaus S (2016) The macromolecular crowding effect – from in vitro into the cell. Biol Chem 397:37–44. https://doi.org/10.1515/hsz-2015-0161
81. Rivas G, Minton AP (2016) Macromolecular crowding in vitro, in vivo, and in between. Trends Biochem Sci 41:970–981. https://doi.org/10.1016/j.tibs.2016.08.013
82. Ellis RJ (2001) Macromolecular crowding: obvious but underappreciated. Trends Biochem Sci 26:597–604. https://doi.org/10.1016/S0968-0004(01)01938-7
83. Cui J, Wu D, Sun Q et al (2020) A PEGDA/DNA hybrid hydrogel for cell-free protein synthesis. Front Chem 8:28. https://doi.org/10.3389/fchem.2020.00028
84. Park N, Kahn JS, Rice EJ et al (2009) High-yield cell-free protein production from P-gel. Nat Protoc 4:1759–1770. https://doi.org/10.1038/nprot.2009.174
85. Park N, Um SH, Funabashi H et al (2009) A cell-free protein-producing gel. Nat Mater 85(8): 432–437. https://doi.org/10.1038/nmat2419
86. Jiao Y, Liu Y, Luo D et al (2018) Microfluidic-assisted fabrication of clay microgels for cell-free protein synthesis. ACS Appl Mater Interfaces 10:29308–29313. https://doi.org/10.1021/acsami.8b09324
87. Oikawa H, Tokiwano T (2004) Enzymatic catalysis of the Diels–Alder reaction in the biosynthesis of natural products. Nat Prod Rep 21:321–352. https://doi.org/10.1039/B305068H

88. Walsh CT, Wencewicz TA (2012) Flavoenzymes: versatile catalysts in biosynthetic pathways. Nat Prod Rep 30:175–200. https://doi.org/10.1039/C2NP20069D
89. Wu X (2022) Multi-enzymatic hydrogel system for biosynthesis. The Chinese University of Hong Kong
90. Aehle W (2007) Enzymes in industry: production and applications.3rd edn. Wiley
91. Rozzell JD (1999) Commercial scale biocatalysis: myths and realities. Bioorg Med Chem 7: 2253–2261. https://doi.org/10.1016/S0968-0896(99)00159-5

Adv Biochem Eng Biotechnol (2023) 186: 77–102
https://doi.org/10.1007/10_2023_221

Published online: 13 June 2023

Compartmentalized Cell-Free Expression Systems for Building Synthetic Cells

Check for updates

David T. Gonzales, Surased Suraritdechachai, and T. -Y. Dora Tang

Contents

1 Introduction 79
1.1 Historical Background 79
1.2 Modern Cell-Free Expression Systems 79
1.3 Applications of CFES in Industry and Health 80
1.4 Purified and Reconstituted Cell-Free Expression Systems 82
2 Compartmentalizing CFES 82
2.1 Lipid Vesicles/Liposomes 83
2.2 Water-in-Oil Emulsions and Droplet Interface Bilayers (DIBs) 84
2.3 Polymersomes 85
3 Biological Models Using Compartmentalized CFES 86
3.1 Cell Growth 86
3.2 Cell Division 88
3.3 Cell Shape Change 88
3.4 Stochastic Gene Expression 89
3.5 Surface Effects in Compartmentalized CFES 89
3.6 Molecular Crowding in the Cytosol 90
3.7 Energy Regeneration for Out-of-Equilibrium Systems 90

D. T. Gonzales
Max Planck Institute of Molecular Cell Biology and Genetics, Dresden, Germany

Center for Systems Biology Dresden, Dresden, Germany

S. Suraritdechachai
Max Planck Institute of Molecular Cell Biology and Genetics, Dresden, Germany

T. -Y. D. Tang (✉)
Max Planck Institute of Molecular Cell Biology and Genetics, Dresden, Germany

Center for Systems Biology Dresden, Dresden, Germany

Physics of Life, Cluster of Excellence, TU Dresden, Dresden, Germany
e-mail: tang@mpi-cbg.de

3.8 Replication 91
3.9 Intercellular Communication and Signal Transduction 92
4 Conclusion 93
References 94

Abstract One of the grand challenges in bottom-up synthetic biology is the design and construction of synthetic cellular systems. One strategy toward this goal is the systematic reconstitution of biological processes using purified or non-living molecular components to recreate specific cellular functions such as metabolism, intercellular communication, signal transduction, and growth and division. Cell-free expression systems (CFES) are in vitro reconstitutions of the transcription and translation machinery found in cells and are a key technology for bottom-up synthetic biology. The open and simplified reaction environment of CFES has helped researchers discover fundamental concepts in the molecular biology of the cell. In recent decades, there has been a drive to encapsulate CFES reactions into cell-like compartments with the aim of building synthetic cells and multicellular systems. In this chapter, we discuss recent progress in compartmentalizing CFES to build simple and minimal models of biological processes that can help provide a better understanding of the process of self-assembly in molecularly complex systems.

Graphical Abstract

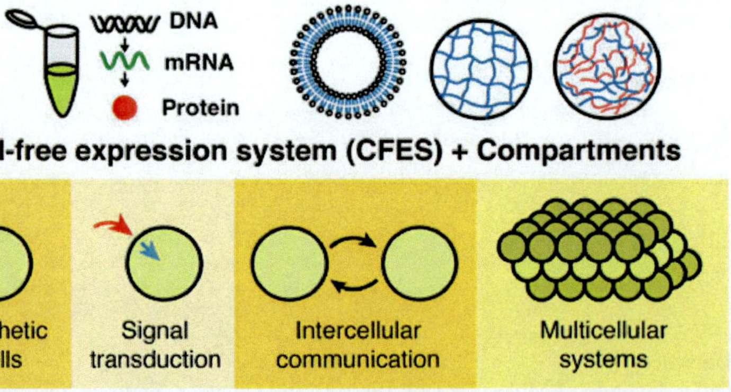

Keywords Cell-free expression systems, Liposomes, Metabolism, Replication, Synthetic cells

1 Introduction

1.1 Historical Background

The systematic reconstitution of biological processes involves purifying and recombining biological components or molecular species to recreate a specific cellular function. This approach was used as early as the late eighteenth century when Eduard Buchner demonstrated that fermentation could still occur in yeast extracts [1]. Although conducted using crude cell extracts, Buchner strengthened the concept that living organisms are driven by a collection of biochemical reactions. Cell extracts provide an open system of the native cellular machinery, making it possible to conveniently investigate the underlying biochemistry in the cell. This led to several important discoveries about ribosome function [2, 3], DNA replication [4, 5], and most notably the genetic coding mechanism [6]. Nirenberg and Matthaei showed how polyuridylic acid RNA templates are translated into poly-L-phenylalanine peptides in *Escherichia coli* extracts – hinting at the translation of the UUU codon to phenylalanine in the genetic code. One of the challenges of using cell extracts was that the extracts had to be prepared fresh for each experiment. In 1961, a protocol was established to prepare *E. coli* cell extracts that could produce protein from RNA [7, 8] and DNA [9, 10] even after storage for several months at −15°C. Extracts prepared were named S-30 and S-100, according to the supernatant fractions obtained from crude *E. coli* cell extracts after centrifugation speeds of 30,000xg and 100,000xg, respectively, and this terminology is still used today to describe cell extracts. The extracts combined with a reaction buffer containing amino acids, NTPs, salts, energy sources, and other cofactors that allow transcription and translation reactions make up what we now refer to as a cell-free expression systems (CFES).

1.2 Modern Cell-Free Expression Systems

In the recent decades, protein yield and ease-of-production of cell-free expression systems have dramatically improved. In *E. coli* cells, protein yield has gone from 1 μg/mL in 1973 [10] to up to 4 mg/L [11] in 2021. Importantly, the use of CFES can be advantageous for producing proteins toxic to the cell or proteins synthesized with unnatural amino acids, both of which can be challenging using in vivo methods. To date, CFES have been developed for many different prokaryotic organisms such as *Vibrio natriegens* [12, 13], *Bacillus subtilis* [14], *Pseudomonas putida* [15], *Streptomyces venezuelae* [16], and *Bacillus megaterium* [17], as well as for eukaryotic organisms such as yeast [18], wheat germ [19], insect [20], HeLa [21], rabbit reticulocyte [22], and plant cells [23, 24]. The different organisms used for extract preparation can provide specific advantages. For example, *Streptomyces*-based extracts are sought because of the capability of *Streptomyces* to express natural product gene clusters and high GC-content genes [16]. The marine bacterium *V. natriegens* has a reported doubling time of 10 min [25] – the fastest growing bacterium known to date. Due to this extremely fast growth rate and potentially enhanced protein translation rates, it has gained interest for applications in recombinant protein expression [26]. However, compared to *E. coli* extract-based CFES, other prokaryotic and eukaryotic CFES generally have lower protein expression yields. Nevertheless, methods for eukaryotic CFES are continually improving and it has been reported that a commercial Tobacco BY-2 CFES can produce up to 3 mg/mL of protein (ALiCE Cell Free Protein Expression Kit, LenioBio). Eukaryotic-based extracts also have a greater capacity to carry out post-translational modifications necessary for many complex proteins and are also suspected to provide a more favorable environment for the proper folding of multi-domain proteins as a result of slower polypeptide elongation rates and co-translational mechanism of chaperones [27]. For example, cell-free expression systems from *Spodoptera frugiperda* 21 (Sf21) insect and Tobacco Bright Yellow 2 (BY-2) plant cell extracts are able to actively glycosylate expressed proteins [24, 28]. To overcome the limitations of *E. coli* extracts, which cannot undertake post-translational modifications, extracts can be pre-enriched with oligosaccharyltransferases and lipid-linked oligosaccharides to glycosylate a cell-free expressed protein target [29, 30]. Disulfide bond formation can also be promoted in *E. coli* extracts by adding glutathione and disulfide isomerase [31]. These examples demonstrate the open nature of cell-free extracts that provides additional flexibility to add new machineries to tune protein structure or incorporate unnatural amino acids [32, 33]. Furthermore, different extracts can even be pooled together to provide diverse regulatory mechanisms within one reaction [34].

1.3 Applications of CFES in Industry and Health

Beyond protein expression, CFES also serve as rapid prototyping platforms that are used to test and optimize genetic circuits [35, 36] and biosynthetic pathways (Fig. 1). For instance, crude extracts enriched with enzymes can be mixed at different ratios to help optimize biosynthetic pathways for n-butanol [37] or mevalonate [38] production. Unlike biological cells, the experimental setup of a cell-free system platform avoids the time-consuming step of DNA transformation and cell culture. However, the industrial use of cell-free reactions to produce molecules of interest is mainly limited by its ability to scale to larger volumes and its prohibitive cost. Low value biomolecules such as subtilisin will not be economically feasible, but high-value pharmaceuticals, such as drug-conjugated antibodies, could be viable as cell-free products [39]. In efforts to improve scale-up performance of CFES, Voloshin and Swartz determined that oxygen availability was crucial for scale-up and demonstrated 1 L cell-free reactions in a stirred tank reactor that produced up to 435 μg/mL of IGF-1 [40]. Zawada et al. produced a cytokine human granulocyte-macrophage colony-stimulating factor (GM-CSF) at 700 μg/mL in a 100 L batch [41] – the largest

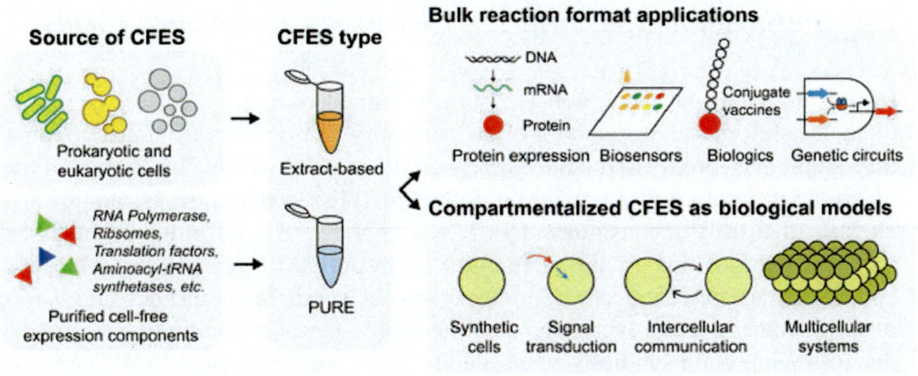

Fig. 1 Cell-free expression systems (CFES) and their applications. CFES can be made from cell extracts or from purified components of the cell (PURE system). As bulk reactions, CFES can be used to rapidly test gene expression DNA constructs to express proteins and genetic circuits or produce biologics such as conjugate vaccines. They are also amenable to freeze-drying in tubes on paper supports and tubes for easy transport biosensors and on-demand manufacturing of medicines. CFES can also be compartmentalized into synthetic cells to act as simplified biological cells and multicellular systems. This provides a platform to study compartmentalized biochemical reactions, signal transduction, intercellular communication, and community effects in a minimal and in vitro context

cell-free batch reaction done. Cell-free systems have also been developed into biosensors for medical, environmental, and forensic diagnostics by coupling nucleic acid sensors or metabolic pathways with transcription factors that trigger a visual output. This has been shown to specifically detect target molecules like RNA sequences from the Norovirus, Zika, and Ebola viruses using RNA toehold switches [42–44] and other substances such as benzoic acid, hippuric acid, and cocaine [45], or Hg(II) and gamma-hydroxybutyrate [46] using novel metabolic enzymes or transcription factors. Lastly, one of the most attractive aspects of cell-free systems is their potential to reduce distribution costs and decentralize production of biomolecules.

CFES are also amenable to freeze-drying, which can remove the cold chain transport and distribution requirements of many biomolecular products. For example, diagnostic kits based on cell-free systems can be freeze-dried onto paper for convenient use in the field [42]. Conjugate vaccines, produced from freeze-dried *E. coli*-based cell-free systems called iVAX (in vitro conjugate vaccine expression), have been shown to be effective against bacterial pathogens in mice [47]. Freeze-dried cell-free diagnostic kits with a glucose output that are compatible with off-the-shelf glucose meters provide a practical approach to low-cost diagnostics [48]. The field of personalized medicine could also potentially use cell-free systems at point-of-care. Extracts from human blood-derived leukocytes have been demonstrated to express nano luciferase (Nluc), granulocyte-colony stimulating factor (G-CSF), and erythropoietin (EPO) [49]. Although many of the applications of cell-free systems mentioned serve as proofs-of-principle to date, new and innovative solutions are continuously being developed to improve yields, specificity, and reproducibility.

1.4 Purified and Reconstituted Cell-Free Expression Systems

Reconstituting gene expression is not only limited to cell extracts. Over the past three decades, our understanding of the necessary components for transcription and translation has progressed and expression systems can now be compiled from individually purified proteins into a defined cell-free expression system. The first attempt of a fully reconstituted CFES was reported by the group of Weissbach (1977). Their purified system, composed of 33 purified *E. coli* factors, ribosomes, and Ehrlich ascites extracts for aminoacyl-tRNA synthetases, did not fully support protein synthesis. However, they identified three partially purified lysate fractions that restored protein synthesis when added to their purified system [50]. These lysate fractions were extracts further purified by 100,000xg centrifugation, ammonium sulfate precipitation, ion exchange, or methanol extraction and suspected to contain other important gene expression components such as the ribosome release factor. Several groups then used different methods to circumvent limitations in the purification of aminoacyl-tRNA synthetases by using pre-charged aminoacyl tRNAs [51] or partially purified aminoacyl-tRNA synthetases [52]. In 2001, the group of Ueda developed a completely defined mix for cell-free expression called the PURE system (protein synthesis using recombinant elements) [53]. The PURE system is composed of 32 His-tagged purified proteins from *E. coli* in addition to tRNAs, NTPs, amino acids, ribosomes, and other components for energy regeneration. The PURE system made it possible to test the effect of individual CFES components on gene expression dynamics and protein yield [54, 55]. However, the production of PURE systems is costly and labor-intensive. This has recently been circumvented by purifying several protein-producing *E. coli* strains in a single co-culture and purification step (OnePot PURE system) to significantly reduce the labor and cost of the PURE CFES [56].

2 Compartmentalizing CFES

Given the wide range of applications of cell-free expression systems, they are ideal as the basis for building synthetic cellular systems by encapsulation into compartments. To this end, CFES (typically *E. coli* extract-based or the PURE system) have been successfully incorporated into or with a broad range of compartments from membrane-bound compartments formed from lipids [57–59], proteins and peptides [60–64], polymersomes [65], inorganic silica nanoparticles [66], clay microgels [67, 68], or metal organic frameworks [69], as well as membrane-free compartments such as coacervates [70, 71] and hydrogels [72–78] (Fig. 2). The latter systems provide synthetic environments which mimic the membrane-free chemically enriched environments of biological condensates and the physically-crowded cytoplasm. In this section, we briefly review several examples of compartmentalized

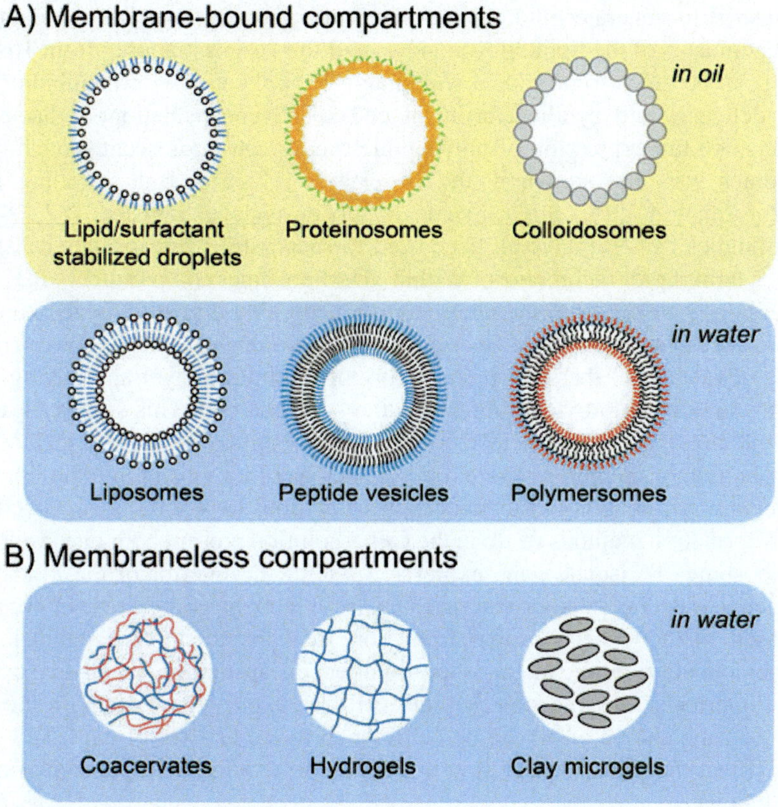

Fig. 2 Examples of compartments that have been reported to encapsulate cell-free expression systems. (**a**) For membrane-bound compartments with a bulk oil phase, CFES have been encapsulated in water-oil emulsions stabilized by surfactants or lipids (droplet interface bilayers, DIBs), crosslinked polymer–protein conjugates (proteinosomes), or silica compartments (colloidosomes). Membrane-bound compartments in aqueous environments encapsulating CFES have been demonstrated in liposomes, amphiphilic peptide vesicles, and polymersomes. (**b**) Membraneless compartments in an aqueous outer environment such as coacervates, hydrogels, or clay microgels can also host the CFES while allowing material exchange with the surroundings

CFES and their properties that should be considered for the design and implementation of bottom-up synthetic cell systems.

2.1 *Lipid Vesicles/Liposomes*

Within the context of synthetic cells, CFES encapsulation into liposomes or lipid vesicles most closely resembles the biological cell. The cell membrane is also composed of a lipid bilayer and is roughly composed of equal proportions of lipids and proteins [79]. Since the 1960s [80, 81], a variety of methods have been established to generate lipid vesicles. Depending on the method, the composition and asymmetry of the lipid bilayer and size of the vesicles ranging from 10s to 100s of nm can be tuned to provide a wide range of applications for membrane science, drug delivery, and synthetic/artificial cells. The encapsulation of bacterial cell extracts and the expression of polypeptides and green fluorescent protein (GFP) in liposomes was first shown in the late 1990s [57, 58]. Bulk swelling [82–84], electroformation [85, 86], inverse emulsion phase transfer [87, 88], and microfluidics [89–91] have all been used to successfully encapsulate cell-free systems within lipid membranes. Within the liposome compartment, the cell-free expression is closed from the outer environment. Consequently, gene expression is limited as resources are depleted as the reaction progresses. This can be overcome by taking advantage of the semi-permeability of the lipid bilayer and having an outer feeding solution to provide nutrients to the synthetic cell. This strategy was shown by Noireaux and Libchaber to increase protein expression from 2 to 5 h. Further incorporation of membrane protein pores such as α-hemolysin into the lipid bilayer to increase permeability extended expression time to 4 days [59]. Swelling and electroformation methods result in the CFES solution present both inside and outside the liposome. To isolate gene expression within the interior of the lipid vesicles, additional washing steps or external DNAse treatment is required. In comparison, oil-based methods of liposome formation, such as the inverse emulsion phase transfer and microfluidic methods lead to direct encapsulation of cell-free expression in the interior of the liposome without cell-free expression system on the outside. However, the choice of oil can be detrimental to the activity of the CFES [92] and could affect the proper integration of membrane proteins due to residual oil in the lipid bilayer.

2.2 *Water-in-Oil Emulsions and Droplet Interface Bilayers (DIBs)*

Although water-in-oil emulsions stabilized by surfactants do not represent the aqueous environments in biology, they offer a stable and convenient method of encapsulation. Unlike liposomes, water-in-oil emulsions do not require an outer solution with a balanced osmolarity relative to the inner CFES. Droplet emulsions can be stabilized by a variety of oil-surfactant combinations such as mineral oil with Span 80 or Tween 80 [93]. Fluorinated oils such as Fluorinert FC-40 (Sigma, USA) or hydrofluoroether oils in combination with 2% w/w EA surfactant (RainDance Technologies, USA) or 2% biocompatible Krytox-based tri-block copolymer surfactant are also typically used due to their biological inertness, capacity for dissolved oxygen, and droplet stability [94]. The choice of surfactant can affect inter-droplet diffusion and material exchange properties. For example, Chowdhury et al. demonstrated that multiple hydrogen bond donors in the surfactant can form a densely connected hydrogen bonding network in the oil–aqueous interface and minimize inter-droplet transfer [94]. Water-in-oil droplets stabilized by lipids and brought in close proximity to each other can form droplet interface bilayers (DIBs) that offer an internal aqueous environment with a lipid bilayer. By encapsulating CFES into DIBs, it has been shown that membrane proteins can be co-translated directly into the lipid bilayer interface. Similar to CFES in liposomes, this provides an environment for simultaneous in situ protein expression and lipid membrane insertion of membrane proteins. For example, Syeda et al. encapsulated commercial CFES (Promega T7-S30 Extract System and Cosmo Bio PURESYSTEM classic II), expressed the viral potassium channel Kcv and α-hemolysin channel in DIBs to connect the aqueous reservoirs [95]. In the same study, it was also noted that the stability of the lipid bilayer interface between connected droplets was reduced when encapsulating the complex CFES mixtures. The PURE encapsulated DIBs lasted an average of 8.7 h while the Promega T7-S30 Extract encapsulated DIBs lasted only 0.7 h. Friddin et al. further investigated the effects of the different components of commercial CFES to DIB stability composed of 4 mg/mL 1:1 (w/w) DOPC/POPG lipids in decane. They found that the extract component of all three commercial CFES results in bilayer failure within 30 min due to their high protein content ($\sim$15 mg/mL). Solutions of bovine serine albumin (BSA), lysozyme, and PEG 8000 were also tested to determine the effects of a positively or negatively charged protein and molecular crowder, respectively, on the stability of the bilayer in DIBs. Only the negatively charged BSA resulted in bilayer failure [96]. Using a lipid mixture composed of 1 mM DPhPC with 10% DPPE-mPEG2000 in 50:50 (v/v) hexadecane/silicone oil, Booth et al. achieved long-term bilayer stability for up to 18 h for PURExpress CFES (NEB, USA) in DIBs [97].

2.3 Polymersomes

Amphiphilic block copolymers, synthetic analogs of the phospholipid molecule, can also form membrane-bound compartments in an aqueous environment called polymersomes. Generally, polymersomes have a lower permeability than liposomes but are more mechanically and chemically stable [98–100]. Given its synthetic nature, membrane properties of polymersomes can be easily modified according to the chemical structure of the amphiphilic block copolymers. For example, the permeability of polymersomes can be tuned by adjusting the thickness and morphology of the membrane [100, 101] or the hydrophobicity and charge of the block polymer relative to the diffusing molecule [102]. Most notably, it has also been shown that membrane proteins can be properly integrated into the polymeric membrane. Polymersomes composed of polybutadiene-poly(ethylene oxide) (PBD-PEO) have been used to encapsulate wheat germ cell extracts to co-translationally express membrane proteins such as α-hemolysin and claudin-2 into the polymeric membrane [103]. The dopamine receptor D2 (DRD2), a G-protein-coupled receptor (GPCR), was also co-translated into the membrane of di- or tri-block copolymer polymersomes. It was also further shown that the integrated GPCRs retained their ligand binding affinity, which suggests proper folding and orientation of the membrane protein [104]. Hybrid lipid-diblock copolymer vesicles can also be made to finely tune lipid membrane properties to enhance membrane protein folding. This was demonstrated using poly(ethylene oxide)-b-polybutadiene (PEO-b-PBD) diblock copolymer and 1,2-dioleoyl-sn-glycero-3-phosphocholine (DOPC) hybrid vesicles, co-expressed with the mechanosensitive membrane protein channel of large conductance tagged with GFP (MscL-GFP). Improvements in protein folding with 10% of the PEO-b-PBD diblock copolymer in the membrane suggested that membrane elasticity plays an important role in membrane protein insertion [105]. Like synthetic block copolymers, elastin-like peptides (ELPs) give an alternative approach to the formation of non-lipid membrane-bound components. ELPs are amphiphilic peptides that are capable of forming vesicles and can be linked directly to the expression machinery of the cell. ELP vesicles or peptidosomes can be formed by swelling methods (179 nm diameter vesicles) [60] or phase transfer methods (15 μm diameter vesicles) [64] with CFES for encapsulated gene expression. These examples show that synthetic and alternative membrane-bound compartments can also support biological processes such as cell-free expression and membrane protein expression, integration, and function.

3 Biological Models Using Compartmentalized CFES

The ability to encapsulate cell-free expression systems into cell-sized compartments opens many possibilities to incorporate cell- and multicellular-like functionalities into synthetic systems (Fig. 3). Although these synthetic systems are chemically very different from what is found in biology, the physical principles governing them remain the same. This provides an abstracted model that is simpler and potentially more experimentally accessible for manipulation relative to biological cells. Here, we highlight the different examples where encapsulated CFES reactions are used to mimic biological processes such as dynamic compartments seen in cell growth and division, maintenance of out-of-equilibrium reactions by energy regeneration, stochastic chemical reactions from low copy number reactants in micron-sized compartments, and intercellular communication and signal transduction.

3.1 Cell Growth

Growth and division is a fundamental property of living systems. Vesicle growth necessitates the incorporation of additional membrane building blocks to allow an increase in cell surface area and volume. The ability for CFES to generate lipids in situ is a significant step toward establishing autonomous synthetic cellular systems that can grow and divide. Danelon et al. (2016, 2020) reconstituted the *E. coli* pathway for the production of phospholipids such as phosphatidic acid (PA),

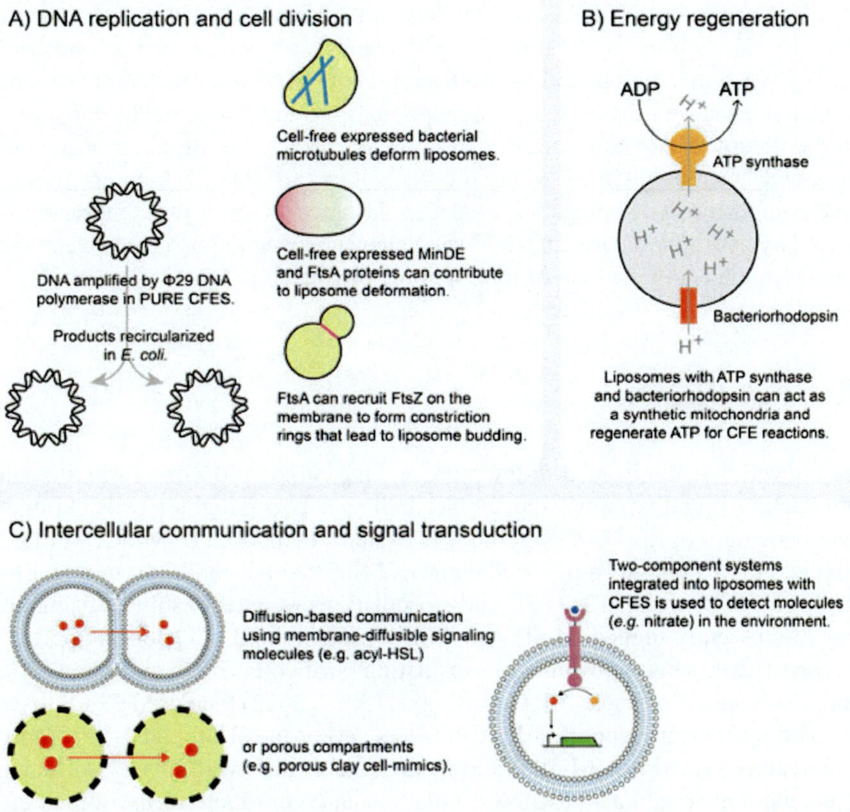

Fig. 3 Some examples of biological processes implemented in compartments driven by CFES. (**a**) DNA replication, cell shape change, and cell division. (**b**) Energy regeneration in liposomes with bacteriorhodopsin to create a proton gradient within the liposome and drive ATP synthase. (**c**) Intercellular communication by a diffusion-mediated signaling molecule and signal transduction by two-component systems

phosphatidylethanolamine (PE), and phosphatidylglycerol (PG) lipids using PURE express on the outside of preformed liposomes [106, 107]. Mass spectrometry measurements indicated that up to 40% of the acyl-CoA can be converted to the phospholipid (or 20 μM of phospholipid end product). Eto et al. later reported a higher yield of phospholipid production of up to 100 μM in liposomes [108]. However, the current yield of biosynthesized lipid is not yet significant enough to affect the size of the liposomes. This could be due to insufficient production of new lipids, changes in internal volume due to osmotic differences, or non-incorporation of the lipids into the membrane. Some of these challenges could be mitigated by increased lipid production using chemoenzymatic pathways [109] and well-defined and controllable experimental settings using microfluidics [110]. Alternative strategies of vesicle growth have been successfully shown with elastin-like peptide (ELP)-based vesicles expressed by CFES encapsulated in ELP vesicles [60]. Synthetic cells can also be made to grow by fusion with other cells. Liposome fusion can be promoted using DNA tethers [111] and SNARE proteins [112]. Using different sequences in DNA tethers allows different liposome populations to be specifically targeted based on the complement sequence. For example, Peruzzi et al. directed fusion of liposomes encapsulating CFES to specific liposomes carrying DNA cargo. In addition, lipid compositions of liposomes were also designed to result in membrane domains caused by hydrophobic mismatch. Fusion events between liposomes are enhanced as it relieves the energetic cost of hydrophobic mismatch [111].

3.2 Cell Division

During cell division in *E. coli*, the proteins FtsZ, FtsA, and ZipA form a membrane constriction ring that splits the mother cell into two daughter cells. Prior to the polymerization of the FtsZ ring, the Min system (composed of MinC, MinD, MinE proteins) drives pole-to-pole oscillations of MinC, a FtsZ inhibitor, to direct FtsZ to the middle of the cell [113–115]. These oscillations were reconstituted using purified and fluorescently-tagged MinE and MinD proteins and ATP on supported lipid bilayers [116], while liposomes encapsulating purified FtsZ proteins showed formation of a contractile ring [117]. Godino et al. (2019, 2022) then used CFES to express the Min system on supported lipid bilayers, water-in-oil droplets, and liposomes. They demonstrated that MinDE and FtsA proteins that were expressed under cell-free conditions could also exhibit oscillations and contribute to liposome deformation [118, 119]. They further showed that cell-free expressed FtsA can recruit purified FtsZ on the membrane in liposomes and form constriction rings that led to liposome budding (without complete scission and release) [120].

3.3 Cell Shape Change

During cell division, active actin filaments and microtubules modify cell shape and play a role in separating chromosomes during cell division. Kattan et al. (2021) expressed bacterial microtubules (bMTs) using CFES in liposomes to drastically deform the liposome shape [121]. This demonstrates that larger cell-free expressed machinery such as bacterial microtubules beyond singular proteins can have active physical functions within liposomes. Although not yet implemented with CFES, other approaches that can modulate liposomes could be integrated with compartmentalized cell-free expression systems. For instance, using phase separated lipids on the lipid membrane or having osmotic imbalances between the inner and outer solutions can be used to trigger liposome division [122]. Membrane proteins can also tune the spontaneous curvature of liposome membranes to generate constriction forces that form membrane necks and complete liposome division [123]. Liposomes can also be stabilized and shaped using programmable DNA elements that self-assemble into cytoskeleton-like structures [124, 125]. Actomyosin rings from purified proteins in giant unilamellar vesicles (GUVs) have been shown to robustly condense into single actin rings and generate constriction forces by ATP-driven myosin motors [126].

3.4 Stochastic Gene Expression

Biological cells are micron-sized compartments which contain low copy numbers of biomolecules. This results in stochastic effects in gene expression [127]. Encapsulating CFES in small cell-sized compartments provides a similar system to mimic stochastic gene expression. Emulsion droplets and liposomes were used to study the effect of cell-sized compartmentalization on low copy numbers and stochasticity of cell-free gene expression. Hansen et al. (2015) generated picoliter droplets (25 μm in diameter or 8.2 pL volume) containing CFES with 100–16,000 copies of DNA plasmid by microfluidics. They showed that, consistent with theory and biological observations [127, 128], noise and variability increase as reactant number decreases and that high levels of crowding can lead to heterogeneous spatial organization of gene expression [129]. Nishimura et al. (2015) used liposomes as cell-sized (1–50 fL) CFES microreactors and found that intrinsic noise levels of these synthetic cells and *E. coli* cells are similar [130].

3.5 Surface Effects in Compartmentalized CFES

Encapsulation of biochemical reaction systems can also be subject to surface effects at the interface of the compartment and outer environment. Wang et al. (2018) encapsulated mammalian-based CFES into water-in-oil emulsion droplets using mineral oil with 2% v/v Span80. They noted differences in transcription and translation dynamics between bulk and droplet-encapsulated reactions [131]. These observations could point to the effects of heterogeneous microenvironments or surface effects that can play a larger role in small volumes. Sakamoto et al. (2018) investigated the role of surface effects in detail using CFES droplets (10–100 μm diameter) in fluorinated oil with 0.1% (w/v) 1,2-dioleoyl-sn-glycero-3-phosphocholine (DOPC) as a surfactant. They showed that expressed deGFP did not scale with droplet volume as geometrically expected ($\propto R^3$). They hypothesized that the large surface to volume ratios in cell-sized droplets can significantly affect gene expression as CFE substrates due to the deactivation (or activation) of mRNA at the surface [132]. In liposome encapsulated CFES, diffusion of materials through the semi-permeable lipid bilayer could affect gene expression conditions and transcription and translation rates [133].

3.6 Molecular Crowding in the Cytosol

The effect of a molecularly crowded environment on liposome encapsulated CFES can provide a model system that mimics the molecularly crowded environment of the cytosol. One approach is to include molecular crowders such as Ficoll-70, an alternative approach is to use the membrane of the liposome to concentrate the CFE mix by osmotic shrinking. Using this method, encapsulated CFES can be concentrated from 14 to 390 mg/mL of total protein. Consequently, 260 mg/mL of deGFP was expressed [134, 135]. Coacervation can also provide a charged and crowded environment. It was shown that the rate and yield of protein expression was affected by the charge and crowded environment that arises from coacervation between two oppositely charged components [70]. Here, mCherry gene expression was enhanced compared to free buffer. However, the coacervate environment could also adversely affect gene expression as protein aggregation within the coacervate phase occurred over time. Overall, molecularly crowded environments are important to consider when studying transcription and translation as these significantly affect protein expression. Not only is this critical for investigating biological function, it can also be readily exploited for application where the crowded environment can be chemically tuned to provide a desired outcome in gene expression.

3.7 Energy Regeneration for Out-of-Equilibrium Systems

Metabolism provides the energy required for the cell to continue functioning in an out-of-equilibrium state. One strategy to provide a continuous energy supply to synthetic cells is by reconstituting the protein motor ATP synthase into liposome synthetic cells. The liposomes provides a hydrophobic environment for the membrane components of the ATP synthase complex, as well as a compartment to store a proton gradient required to drive ATP synthase.

Matthies et al. (2011) showed that proteoliposomes mixed with CFES expressing the *atp*-operon resulted in completely assembled ATP synthase complexes on the lipid membrane [136]. However, functional studies of the ATP synthase integrated proteoliposomes were not performed. Berhanu et al. (2019) paired purified ATP synthase and bacterial rhodopsin in proteoliposomes. Driven by light, the bacteriorhodopsin generates a proton gradient in the proteoliposome, which in turn powers the ATP synthase to produce ATP from ADP. This artificial organelle was encapsulated within GUVs with CFES to provide ATP for cell-free expression [137]. Another strategy is to immobilize the CFES in permeable hydrogel particles and continuously supply nutrients and energy by flow. Not only does this strategy allow replenishment of energy in the CFES, it also provides a mechanism of turnover that is essential to out-of-equilibrium systems. It has been shown that cell-free gene expression is compatible with a variety of hydrogels based on agarose, agar, xanthan, and polyacrylamide [72]. The chemical and structural characteristics of the different hydrogels, which can be composed of entangled polymers, micellar aggregates, or covalently crosslinked polymers, can provide unique properties for carrying CFE reactions. Different hydrogels can offer varying degrees of molecular crowding and porosity to promote cell-free expression. However, due to the open nature of hydrogel particles, it can be challenging to spatially localize the CFES components as DNA and proteins can easily diffuse out of the hydrogel. To avoid this, the chemistry of hydrogels can be tuned and modified to covalently fix certain CFES components, such as DNA, into the hydrogel matrix and avoid loss of the molecule during nutrient flow. For example, chemical modifications to hyaluronic acid with thiols [73] or 5-methylfuran and then dibenzocyclooctyne on the carboxyl groups [74] provide sites for covalently linking modified DNA templates to localize gene expression within the hydrogel. Zhou et al. (2018) used a Ni^{2+}-NTA-functionalized polyacrylamide to immobilize the His-tagged components of the PURE CFES into hydrogel beads. By continually supplying the CFES-loaded hydrogel with nutrients, the PURE CFES was able to support gene expression for up to 28 days [77].

3.8 Replication

Replication is an essential biological process that copies double-stranded DNA into two identical copies which are then propagated to daughter cells when the cell divides. Transcription–translation-coupled DNA replication (TTcDR) systems using CFES have been successfully implemented to DNA replication. This was done by expressing the Φ29 DNA polymerase for rolling circle amplification (RCA) of the DNA plasmid template and then circularization of linear strands via the two loxP sites on the DNA using the Cre recombinase. This process was encapsulated into water-in-oil droplets to facilitate in vitro evolution experiments to search for DNA mutations that avoid inhibitory effects of the Cre recombinase to DNA polymerization [138, 139]. Van Nies et al. (2018) further attempted to replicate DNA using a minimal set of two expressed genes, the DNA polymerase (DNAP) and terminal protein (TP), and two externally added purified proteins, the single-stranded DNA binding protein (SSB) and double-stranded DNA binding protein (DSB), from the Φ29 virus with CFES in liposomes [140]. Libicher et al. (2020) optimized the PURExpress (NEB, USA) system to account for the observation that high tRNA and rNTP concentrations reduce the activity of DNA polymerase activity [141]. This modified PURE system called PURErep enabled efficient TTcDR from only Φ29 DNAP without additional replication proteins and yielded concatemer RCA products which can be directly transformed into *E. coli* to recircularize via in vivo intramolecular homologous recombination [142]. The PURErep system was not yet tested in liposomes but is able to replicate a multipartite genome of more than 116 kb as well as express 30 encoded transcription factors in bulk reactions. Self-replication of genomic RNA has also been demonstrated within lipid vesicles [143] and aqueous two-phase systems formed from PEG and Dextran [144]. For the latter, the RNA template and the cell-free translation machinery partition into the Dextran-rich droplets to support RNA self-replication by the translation of the core subunit of Qβ replicase. It was further shown that the Dextran-rich droplets can protect the genomic RNA from short parasitic RNAs by segregating them into distinct droplets. The synthesis of ribosomes is also a crucial step in the construction of a minimal cell. The in vitro integrated synthesis, assembly, and translation (iSAT) method reconstitutes functional ribosomes from ribosome-free extracts by transcribing rRNA and assembling them with ribosomal proteins into functional ribosomes [145]. This system was encapsulated into liposomes to demonstrate its functionality in cell-like compartments [146].

3.9 Intercellular Communication and Signal Transduction

Intercellular communication allows single cells to coordinate into multicellular systems. In synthetic cells, intercellular communication is typically implemented via diffusion of a signaling molecule from one cell to another. Depending on the properties of the compartment, different CFES products can be used as signaling molecules. For example, Adamala et al. (2017) made two populations of liposomes that communicate via the inducer molecule isopropyl-β-D-thiogalactoside (IPTG), which is only released by the sender liposome after expression of α-hemolysin pore proteins [112]. Niederholtmeyer et al. (2018) encapsulated CFE reactions in porous clay artificial cell mimics with a nucleus-like DNA hydrogel compartment. Expressed proteins such as T3 RNA polymerase was able to diffuse freely through the porous clay compartments to activate gene expression in other cells [147]. Dupin et al. (2019, 2022) used multicellular structures of DIBs with extract-based CFES that communicate via membrane diffusible molecules such as N-(3-oxohexanoyl)-L-homoserine lactone (3OC6-HSL) and 3,5-difluoro-4-hydroxybenzylidene imidazolinone (DFHBI), or IPTG diffusing through the droplet bilayer interface via α-hemolysin pore proteins [148, 149]. As an alternative to diffusion-mediated communication, synthetic cells can be designed with signal transducing membrane proteins to relay an extracellular signal into the cell. Biological cells predominantly interact with the environment and other cells through signal transduction. For example, signal transduction in bacteria can occur through one or two-components systems and extracytoplasmic function sigma and anti-sigma factors, all of which have a transmembrane protein component [150]. However, efficient integration of transmembrane proteins can be challenging. One advantage of using CFES with lipid membrane-based compartments is that membrane proteins can be co-translated directly into a lipid environment. However, the effects of lipid environment with respect to proper protein folding are still not fully understood. Jacobs et al. (2019) studied the effect of membrane composition and properties on protein folding of the mechanosensitive channel of large conductance (MscL) channel protein. Using a variety of lipid membrane additives (e.g., diblock copolymers, $C_{12}E_8$ detergent, cholesterol, and PEG), they found that co-translation and proper folding of the

MscL protein can be significantly improved depending on membrane elasticity and the available membrane surface area [105]. Peruzzi et al. investigated how the hydrophobic properties between the lipid membrane and membrane protein can affect protein expression and co-translational folding in a CFES environment. Using de novo designed membrane proteins and liposomes composed of lipids of different fatty acid tail lengths, they were able to systematically show that reducing membrane-protein hydrophobic mismatch can improve expression and integration of membrane proteins [151]. Steinkühler et al. (2023) identified that stalling of the ribosomes and aggregation of the nascent protein chains inhibit CFE of membrane proteins and that providing a lipid membrane environment with the right properties can help alleviate aggregation and promote co-translation of the membrane protein [152]. Further recent work by Peruzzi et al. incorporated the NarX-NarL two-component signal transduction system in liposomes to detect nitrate in the outer environment [153]. In this system, NarX is the co-translated membrane-bound histidine kinase sensor protein that detects nitrate and phosphorylates the cytosolic response regulator NarL for downstream gene expression. By facilitating the use of signal transduction systems and transmembrane proteins in bottom-up synthetic multicellular systems, we can begin to access a rich diversity of extracellular and intercellular functions.

4 Conclusion

In this chapter, we provided a brief historical overview of cell-free expression systems and discuss their utility for building synthetic cellular systems from scratch. In particular, we focused on the use of micron-sized encapsulation of CFES for establishing different models of biological processes in synthetic cells. Over the last decade, much progress has been made in utilizing cell-free systems for reconstituting specific cellular functions in an in vitro environment and simplified context. However, there are still some challenges in compartmentalized CFES that limits our capability to mimic life-like functions. For example, energy regeneration, turnover, and degradation are key limitations in CFES, especially within a compartmentalized system, to maintain out-of-equilibrium reactions. Implementing DNA replication *and* cell division together to achieve self-replication will further require the integration of compatible biochemical and mechanical processes. Therefore, exploring new materials and functions in CFES is important as it will provide more flexibility and options to build more complex and compatible systems. Advancements in the field of bottom-up synthetic biology also offer opportunities to diversify the applications of encapsulated CFES beyond building synthetic cell. For example, light-activated systems in encapsulated CFES [97, 154] could be embedded within tissues and spatially control engineered processes in nearby cells [155]. Synthetic cells could also act as an adaptive drug delivery system that is capable of producing an active protein in situ [156] or sensing and appropriately responding to the nearby environment [157]. Overall, compartmentalized CFES can provide minimal models of biological processes that can help us study the process of self-assembly in molecularly complex systems, as well as potential applications in health and medicine through advanced drug delivery systems.

References

1. Buchner E (1897) Alkoholische Gährung ohne Hefezellen. Ber Dtsch Chem Ges 30:117–124
2. Hoagland MB, Zamecnik PC, Stephenson ML (1957) Intermediate reactions in protein biosynthesis. Biochim Biophys Acta 24:215–216
3. Kirsch JF, Siekevitz P, Palade GE (1960) Amino acid incorporation in vitro by ribonucleoprotein particles detached from Guinea pig liver microsomes. J Biol Chem 235:1419–1424
4. Fuller RS, Kaguni JM, Kornberg A (1981) Enzymatic replication of the origin of the Escherichia coli chromosome. Proc Natl Acad Sci 78:7370–7374
5. Blow JJ, Laskey RA (1986) Initiation of DNA replication in nuclei and purified DNA by a cell-free extract of xenopus eggs. Cell 47:577–587
6. Nirenberg MW, Matthaei JH (1961) The dependence of cell-free protein synthesis in E. coli upon naturally occurring or synthetic polyribonucleotides. Proc Natl Acad Sci 47:1588–1602
7. Matthaei JH, Nirenberg MW (1961) Characteristics and stabilization of DNAase-sensitive protein synthesis in E. coli extracts. Proc Natl Acad Sci 47:1580–1588
8. Nathans D, Notani G, Schwartz JH, Zinder ND (1962) Biosynthesis of the coat protein of coliphage f2 by E. coli extracts. Proc Natl Acad Sci 48:1424–1431
9. DeVries JK, Zubay G (1967) DNA-directed peptide synthesis. II. The synthesis of the alpha-fragment of the enzyme beta-galactosidase. Proc Natl Acad Sci 57:1010–1012
10. Zubay G (1973) In vitro synthesis of protein in microbial systems. Annu Rev Genet 7:267–287
11. Garenne D, Thompson S, Brisson A, Khakimzhan A, Noireaux V (2021) The all- E. coli TXTL toolbox 3.0: new capabilities of a cell-free synthetic biology platform. Synth Biol 5: 344–355
12. Des Soye BJ, Davidson SR, Weinstock MT, Gibson DG, Jewett MC (2018) Establishing a high-yielding cell-free protein synthesis platform derived from vibrio natriegens. ACS Synth Biol 7:2245–2255
13. Wiegand DJ, Lee HH, Ostrov N, Church GM (2019) Cell-free protein expression using the rapidly growing bacterium Vibrio natriegens. J Vis Exp. https://doi.org/10.3791/59495
14. Kelwick R, Webb AJ, MacDonald JT, Freemont PS (2016) Development of a Bacillus subtilis cell-free transcription-translation system for prototyping regulatory elements. Metab Eng 38: 370–381
15. Wang H, Li J, Jewett MC (2018) Development of a Pseudomonas putida cell-free protein synthesis platform for rapid screening of gene regulatory elements. Synth Biol 3:1–7
16. Li J, Wang H, Kwon YC, Jewett MC (2017) Establishing a high yielding streptomyces-based cell-free protein synthesis system. Biotechnol Bioeng 114:1343–1353
17. Moore SJ, MacDonald JT, Wienecke S, Ishwarbhai A, Tsipa A, Aw R, Kylilis N, Bell DJ, McClymont DW, Jensen K et al (2018) Rapid acquisition and model-based analysis of cell-free transcription–translation reactions from nonmodel bacteria. Proc Natl Acad Sci 115: E4340–E4349
18. Hodgman CE, Jewett MC (2013) Optimized extract preparation methods and reaction conditions for improved yeast cell-free protein synthesis. Biotechnol Bioeng 110:2643–2654
19. Takai K, Sawasaki T, Endo Y (2010) Practical cell-free protein synthesis system using purified wheat embryos. Nat Protoc 5:227–238
20. Ezore T, Suzuki T, Higashide S, Shintani E, Endo K, Kobayashi SI, Shikata M, Ito M, Tanimizu K, Nishimura O (2006) Cell-free protein synthesis system prepared from insect cells by freeze-thawing. Biotechnol Prog 22:1570–1577

21. Mikami S, Kobayashi T, Masutani M, Yokoyama S, Imataka H (2008) A human cell-derived in vitro coupled transcription/translation system optimized for production of recombinant proteins. Protein Expr Purif 62:190–198
22. Stavnezer J, Huang RCC (1971) Synthesis of a mouse immunoglobulin light chain in a rabbit reticulocyte cell-free system. Nat New Biol 230:172–176
23. Buntru M, Vogel S, Spiegel H, Schillberg S (2014) Tobacco BY-2 cell-free lysate: an alternative and highly-productive plant-based in vitro translation system. BMC Biotechnol 14:1–11
24. Buntru M, Vogel S, Stoff K, Spiegel H, Schillberg S (2015) A versatile coupled cell-free transcription-translation system based on tobacco BY-2 cell lysates. Biotechnol Bioeng 112: 867–878
25. Eagon R (1962) Pseudomonas natriegens, a marine bacterium with a generation time of less than 10 minutes. J Bacteriol 83:736–737
26. Weinstock MT, Hesek ED, Wilson CM, Gibson DG (2016) Vibrio natriegens as a fast-growing host for molecular biology. Nat Methods 13:849–851
27. Hirano N, Sawasaki T, Tozawa Y, Endo Y, Takai K (2006) Tolerance for random recombination of domains in prokaryotic and eukaryotic translation systems: limited interdomain misfolding in a eukaryotic translation system. Proteins Struct Funct Bioinforma 64:343–354
28. Tarui H, Imanishi S, Hara T (2000) A novel cell-free translation/glycosylation system prepared from insect cells. J Biosci Bioeng 90:508–514
29. Jaroentomeechai T, Stark JC, Natarajan A, Glasscock CJ, Yates LE, Hsu KJ, Mrksich M, Jewett MC, Delisa MP (2018) Single-pot glycoprotein biosynthesis using a cell-free transcription-translation system enriched with glycosylation machinery. Nat Commun 9:1–11
30. Kightlinger W, Duncker KE, Ramesh A, Thames AH, Natarajan A, Stark JC, Yang A, Lin L, Mrksich M, DeLisa MP et al (2019) A cell-free biosynthesis platform for modular construction of protein glycosylation pathways. Nat Commun 10
31. Matsuda T, Watanabe S, Kigawa T (2013) Cell-free synthesis system suitable for disulfide-containing proteins. Biochem Biophys Res Commun 431:296–301
32. Goerke AR, Swartz JR (2009) High-level cell-free synthesis yields of proteins containing site-specific non-natural amino acids. Biotechnol Bioeng 102:400–416
33. Des Soye BJ, Patel JR, Isaacs FJ, Jewett MC (2015) Repurposing the translation apparatus for synthetic biology. Curr Opin Chem Biol 28:83–90
34. Yim SS, Johns NI, Park J, Gomes AL, McBee RM, Richardson M, Ronda C, Chen SP, Garenne D, Noireaux V et al (2019) Multiplex transcriptional characterizations across diverse bacterial species using cell-free systems. Mol Syst Biol 15:1–15
35. Chappell J, Jensen K, Freemont PS (2013) Validation of an entirely in vitro approach for rapid prototyping of DNA regulatory elements for synthetic biology. Nucleic Acids Res 41:3471–3481
36. Niederholtmeyer H, Sun ZZ, Hori Y, Yeung E, Verpoorte A, Murray RM, Maerkl SJ (2015) Rapid cell-free forward engineering of novel genetic ring oscillators. elife 4:1–18
37. Karim AS, Jewett MC (2016) A cell-free framework for rapid biosynthetic pathway prototyping and enzyme discovery. Metab Eng 36:116–126
38. Dudley QM, Anderson KC, Jewett MC (2016) Cell-free mixing of Escherichia coli crude extracts to prototype and rationally engineer high-titer mevalonate synthesis. ACS Synth Biol 5:1578–1588
39. Silverman AD, Karim AS, Jewett MC (2020) Cell-free gene expression: an expanded repertoire of applications. Nat Rev Genet 21:151–170
40. Voloshin AM, Swartz JR (2008) Large-scale batch reactions for cell-free protein synthesis. In: Cell-free protein synthesis
41. Zawada JF, Yin G, Steiner AR, Yang J, Naresh A, Roy SM, Gold DS, Heinsohn HG, Murray CJ (2011) Microscale to manufacturing scale-up of cell-free cytokine production-a new approach for shortening protein production development timelines. Biotechnol Bioeng 108: 1570–1578

42. Pardee K, Green AA, Ferrante T, Cameron DE, Daleykeyser A, Yin P, Collins JJ (2014) Paper-based synthetic gene networks. Cell 159:940–954
43. Pardee K, Green AA, Takahashi MK, Braff D, Lambert G, Lee JW, Ferrante T, Ma D, Donghia N, Fan M et al (2016) Rapid, low-cost detection of zika virus using programmable biomolecular components. Cell 165:1255–1266
44. Ma D, Shen L, Wu K, Diehnelt CW, Green AA (2018) Low-cost detection of norovirus using paper-based cell-free systems and synbody-based viral enrichment. Synth Biol 3:1–11
45. Voyvodic PL, Pandi A, Koch M, Conejero I, Valjent E, Courtet P, Renard E, Faulon JL, Bonnet J (2019) Plug-and-play metabolic transducers expand the chemical detection space of cell-free biosensors. Nat Commun 10:1–8
46. Gräwe A, Dreyer A, Vornholt T, Barteczko U, Buchholz L, Drews G, Ho UL, Jackowski ME, Kracht M, Lüders J et al (2019) A paper-based, cell-free biosensor system for the detection of heavy metals and date rape drugs. PLoS One 14:1–22
47. Stark JC, Jaroentomeechai T, Moeller TD, Hershewe JM, Warfel KF, Moricz BS, Martini AM, Dubner RS, Hsu KJ, Stevenson TC et al (2021) On-demand biomanufacturing of protective conjugate vaccines. Sci Adv 7
48. Amalfitano E, Karlikow M, Norouzi M, Jaenes K, Cicek S, Masum F, Sadat Mousavi P, Guo Y, Tang L, Sydor A et al (2021) A glucose meter interface for point-of-care gene circuit-based diagnostics. Nat Commun 12:1–10
49. Burgenson D, Gurramkonda C, Pilli M, Ge X, Andar A, Kostov Y, Tolosa L, Rao G (2018) Rapid recombinant protein expression in cell-free extracts from human blood. Sci Rep 8:1–7
50. Kung H, Redfield B, Treadwell BV, Eskin B, Spears C, Weissbach H (1977) DNA-directed in vitro synthesis of β-galactosidase. Studies with purified factors. J Biol Chem 252:6889–6894
51. Ganoza MC, Cunningham C, Green RM (1985) Isolation and point of action of a factor from Escherichia coli required to reconstruct translation. Proc Natl Acad Sci U S A 82:1648–1652
52. Pavlov MY, Ehrenberg M (1996) Rate of translation of natural mRNAs in an optimized in vitro system. Arch Biochem Biophys 328:9–16
53. Shimizu Y, Inoue A, Tomari Y, Suzuki T, Yokogawa T, Nishikawa K, Ueda T (2001) Cell-free translation reconstituted with purified components. Nat Biotechnol 19:751–755
54. Matsuura T, Kazuta Y, Aita T, Adachi J, Yomo T (2009) Quantifying epistatic interactions among the components constituting the protein translation system. Mol Syst Biol 5:1–10
55. Doerr A, De Reus E, Van Nies P, Van Der Haar M, Wei K, Kattan J, Wahl A, Danelon C (2019) Modelling cell-free RNA and protein synthesis with minimal systems. Phys Biol 16
56. Lavickova B, Maerkl SJ (2019) A simple, robust, and low-cost method to produce the PURE cell-free system. ACS Synth Biol 8:455–462
57. Oberholzer T, Nierhaus KH, Luisi PL (1999) Protein expression in liposomes. Biochem Biophys Res Commun 261:238–241
58. Yu W, Sato K, Wakabayashi M, Nakaishi T, Ko-Mitamura EP, Shima Y, Urabe I, Yomo T (2001) Synthesis of functional protein in liposome. J Biosci Bioeng 92:590–593
59. Noireaux V, Libchaber A (2004) A vesicle bioreactor as a step toward an artificial cell assembly. Proc Natl Acad Sci 101:17669–17674
60. Vogele K, Frank T, Gasser L, Goetzfried MA, Hackl MW, Sieber SA, Simmel FC, Pirzer T (2018) Towards synthetic cells using peptide-based reaction compartments. Nat Commun 9:1–7
61. Frank T, Vogele K, Dupin A, Simmel FC, Pirzer T (2020) Growth of giant peptide vesicles driven by compartmentalized transcription–translation activity. Chem – A Eur J 26:17356–17360
62. Schreiber A, Huber MC, Schiller SM (2019) Prebiotic protocell model based on dynamic protein membranes accommodating anabolic reactions. Langmuir 35:9593–9610
63. Huang X, Li M, Green DC, Williams DS, Patil AJ, Mann S (2013) Interfacial assembly of protein–polymer nano-conjugates into stimulus-responsive biomimetic protocells. Nat Commun 4:1–9

64. Sharma B, Ma Y, Hiraki HL, Baker BM, Ferguson AL, Liu AP (2021) Facile formation of giant elastin-like polypeptide vesicles as synthetic cells. Chem Commun 57:13202–13205
65. Martino C, Kim SH, Horsfall L, Abbaspourrad A, Rosser SJ, Cooper J, Weitz DA (2012) Protein expression, aggregation, and triggered release from polymersomes as artificial cell-like structures. Angew Chem Int Ed 51:6416–6420
66. Li M, Green DC, Anderson JLR, Binks BP, Mann S (2011) In vitro gene expression and enzyme catalysis in bio-inorganic protocells. Chem Sci 2:1739
67. Jiao Y, Liu Y, Luo D, Huck WTS, Yang D (2018) Microfluidic-assisted fabrication of clay microgels for cell-free protein synthesis. ACS Appl Mater Interfaces 10:29308–29313
68. Yang D, Peng S, Hartman MR, Gupton-Campolongo T, Rice EJ, Chang AK, Gu Z, Lu GQ, Luo D (2013) Enhanced transcription and translation in clay hydrogel and implications for early life evolution. Sci Rep 3:1–6
69. Liu J, Guo Z, Li Y, Liang J, Xue J, Xu J, Whitelock JM, Xie L, Kong B, Liang K (2021) pH-gated activation of gene transcription and translation in biocatalytic metal–organic framework artificial cells. Adv NanoBiomed Res 1:2000034
70. Tang T-YD, van Swaay D, DeMello A, Ross Anderson JL, Mann S (2015) In vitro gene expression within membrane-free coacervate protocells. Chem Commun 51:11429–11432
71. Deng N, Huck WTS (2017) Microfluidic formation of monodisperse coacervate organelles in liposomes. Angew Chem 129:9868–9872
72. Whitfield CJ, Banks AM, Dura G, Love J, Fieldsend JE, Goodchild SA, Fulton DA, Howard TP (2020) Cell-free protein synthesis in hydrogel materials. Chem Commun 56:7108–7111
73. Thiele J, Ma Y, Foschepoth D, Hansen MMK, Steffen C, Heus HA, Huck WTS (2014) DNA-functionalized hydrogels for confined membrane-free in vitro transcription/translation. Lab Chip 14:2651
74. Heida T, Köhler T, Kaufmann A, Männel MJ, Thiele J (2020) Cell-free protein synthesis in bifunctional hyaluronan microgels: a strategy for in situ immobilization and purification of his-tagged proteins. ChemSystemsChem 2:1–7
75. Köhler T, Heida T, Hoefgen S, Weigel N, Valiante V, Thiele J (2020) Cell-free protein synthesis and: in situ immobilization of deGFP-MatB in polymer microgels for malonate-to-malonyl CoA conversion. RSC Adv 10:40588–40596
76. Kahn JS, Ruiz RCH, Sureka S, Peng S, Derrien TL, An D, Luo D (2016) DNA microgels as a platform for cell-free protein expression and display. Biomacromolecules 17:2019–2026
77. Zhou X, Wu H, Cui M, Lai SN, Zheng B (2018) Long-lived protein expression in hydrogel particles: towards artificial cells. Chem Sci 9:4275–4279
78. Lai SN, Zhou X, Ouyang X, Zhou H, Liang Y, Xia J, Zheng B (2020) Artificial cells capable of long-lived protein synthesis by using aptamer grafted polymer hydrogel. ACS Synth Biol 9: 76–83
79. Strahl H, Errington J (2017) Bacterial membranes: structure, domains, and function. Annu Rev Microbiol 71:519–538
80. De Gier J, Mandersloot JG, Van Deenen LLM (1968) Lipid composition and permeability of liposomes. Biochim Biophys Acta Biomembr 150:666–675
81. Reeves JP, Dowben RM (1969) Formation and properties of thin-walled phospholipid vesicles. J Cell Physiol 73:49–60
82. Horger KS, Estes DJ, Capone R, Mayer M (2009) Films of agarose enable rapid formation of giant liposomes in solutions of physiologic ionic strength. J Am Chem Soc 131:1810–1819
83. Weinberger A, Tsai FC, Koenderink GH, Schmidt TF, Itri R, Meier W, Schmatko T, Schröder A, Marques C (2013) Gel-assisted formation of giant unilamellar vesicles. Biophys J 105:154–164
84. Tsumoto K, Matsuo H, Tomita M, Yoshimura T (2009) Efficient formation of giant liposomes through the gentle hydration of phosphatidylcholine films doped with sugar. Colloids Surf B Biointerfaces 68:98–105
85. Taylor P, Xu C, Fletcher PDI, Paunov VN (2003) A novel technique for preparation of monodisperse giant liposomes. Chem Commun 3:1732

86. Pott T, Bouvrais H, Méléard P (2008) Giant unilamellar vesicle formation under physiologically relevant conditions. Chem Phys Lipids 154:115–119
87. Moga A, Yandrapalli N, Dimova R, Robinson T (2019) Optimization of the inverted emulsion method for high-yield production of biomimetic giant unilamellar vesicles. Chembiochem 20: 2674–2682
88. Pautot S, Frisken BJ, Weitz DA (2003) Production of unilamellar vesicles using an inverted emulsion. Langmuir 19:2870–2879
89. Van de Cauter L, Fanalista F, van Buren L, De Franceschi N, Godino E, Bouw S, Danelon C, Dekker C, Koenderink GH, Ganzinger KA (2021) Optimized cDICE for efficient reconstitution of biological systems in giant unilamellar vesicles. ACS Synth Biol 10:1690–1702
90. Deshpande S, Caspi Y, Meijering AEC, Dekker C (2016) Octanol-assisted liposome assembly on chip. Nat Commun 7:1–9
91. Yandrapalli N, Petit J, Bäumchen O, Robinson T (2020) Surfactant-free production of biomimetic artificial cells using PDMS-based microfluidics. bioRxiv. https://doi.org/10.1101/2020.10.23.346932
92. Kaltenbach M, Devenish SRA, Hollfelder F (2012) A simple method to evaluate the biochemical compatibility of oil/surfactant mixtures for experiments in microdroplets. Lab Chip 12: 4185–4192
93. Griffiths AD, Tawfik DS (1998) Man-made cell-like compartments for molecular evolution. Nat Biotechnol 16:652–656
94. Chowdhury MS, Zheng W, Kumari S, Heyman J, Zhang X, Dey P, Weitz DA, Haag R (2019) Dendronized fluorosurfactant for highly stable water-in-fluorinated oil emulsions with minimal inter-droplet transfer of small molecules. Nat Commun 10
95. Syeda R, Holden MA, Hwang WL, Bayley H (2008) Screening blockers against a potassium channel with a droplet interface bilayer array. J Am Chem Soc 130:15543–15548
96. Friddin MS, Morgan H, de Planque MRR (2013) Cell-free protein expression systems in microdroplets: stabilization of interdroplet bilayers. Biomicrofluidics 7
97. Booth MJ, Schild VR, Graham AD, Olof SN, Bayley H (2016) Light-activated communication in synthetic tissues. Sci Adv 2:1–12
98. Rideau E, Dimova R, Schwille P, Wurm FR, Landfester K (2018) Liposomes and polymersomes: a comparative review towards cell mimicking. Chem Soc Rev 47:8572–8610
99. Kamat NP, Katz JS, Hammer DA (2011) Engineering polymersome protocells. J Phys Chem Lett 2:1612–1623
100. Discher BM, Won Y, Ege DS, Lee JC, Bates FS, Discher DE, Hammer DA (1999) Polymersomes: tough vesicles made from diblock copolymers. Science 284:1143–1146
101. Battaglia G, Ryan AJ (2005) Bilayers and interdigitation in block copolymer vesicles. J Am Chem Soc 127:8757–8764
102. So S, Lodge TP (2014) Rate of molecular exchange through the membranes of ionic liquid filled polymersomes dispersed in water. J Phys Chem C 118:21140–21147
103. Nallani M, Andreasson-Ochsner M, Tan C-WD, Sinner E-K, Wisantoso Y, Geifman-Shochat-S, Hunziker W (2011) Proteopolymersomes: in vitro production of a membrane protein in polymersome membranes. Biointerphases 6:153
104. May S, Andreasson-Ochsner M, Fu Z, Low YX, Tan D, De Hoog HPM, Ritz S, Nallani M, Sinner EK (2013) In vitro expressed GPCR inserted in polymersome membranes for ligand-binding studies. Angew Chem Int Ed 52:749–753
105. Jacobs ML, Boyd MA, Kamat NP (2019) Diblock copolymers enhance folding of a mechanosensitive membrane protein during cell-free expression. Proc Natl Acad Sci U S A 116:4031–4036
106. Blanken D, Foschepoth D, Serrão AC, Danelon C (2020) Genetically controlled membrane synthesis in liposomes. Nat Commun 11:1–13
107. Scott A, Noga MJ, De Graaf P, Westerlaken I, Yildirim E, Danelon C (2016) Cell-free phospholipid biosynthesis by gene-encoded enzymes reconstituted in liposomes. PLoS One 11:1–23

108. Eto S, Matsumura R, Shimane Y, Fujimi M, Berhanu S, Kasama T, Kuruma Y (2022) Phospholipid synthesis inside phospholipid membrane vesicles. Commun Biol 5:1–11
109. Bhattacharya A, Cho CJ, Brea RJ, Devaraj NK (2021) Expression of fatty acyl-CoA ligase drives one-pot de novo synthesis of membrane-bound vesicles in a cell-free transcription-translation system. J Am Chem Soc 143:11235–11242
110. Robinson T, Kuhn P, Eyer K, Dittrich PS (2013) Microfluidic trapping of giant unilamellar vesicles to study transport through a membrane pore. Biomicrofluidics 7:044105
111. Peruzzi JA, Jacobs ML, Vu TQ, Wang KS, Kamat NP (2019) Barcoding biological reactions with DNA-functionalized vesicles. Angew Chem 131:18856–18863
112. Adamala KP, Martin-Alarcon DA, Guthrie-Honea KR, Boyden ES (2017) Engineering genetic circuit interactions within and between synthetic minimal cells. Nat Chem 9:431–439
113. De Boer PAJ, Crossley RE, Rothfield LI (1990) Central role for the Escherichia coli minC gene product in two different cell division-inhibition systems. Proc Natl Acad Sci U S A 87: 1129–1133
114. Hu Z, Mukherjee A, Pichoff S, Lutkenhaus J (1999) The MinC component of the division site selection system in Escherichia coli interacts with FtsZ to prevent polymerization. Proc Natl Acad Sci U S A 96:14819–14824
115. De Boer PAJ, Crossley RE, Rothfield LI (1992) Roles of MinC and MinD in the site-specific septation block mediated by the MinCDE system of Escherichia coli. J Bacteriol 174:63–70
116. Loose M, Fischer-Friedrich E, Ries J, Kruse K, Schwille P (2008) Spatial regulators for bacterial cell division self-organize into surface waves in vitro. Science 320:789–792
117. Osawa M, Anderson DE, Erickson HP (2008) Reconstitution of contractile FtsZ rings in liposomes. Science 320:792–794
118. Godino E, López JN, Foschepoth D, Cleij C, Doerr A, Castellà CF, Danelon C (2019) De novo synthesized min proteins drive oscillatory liposome deformation and regulate FtsA-FtsZ cytoskeletal patterns. Nat Commun 10:4969
119. Godino E, Doerr A, Danelon C (2022) Min waves without MinC can pattern FtsA-anchored FtsZ filaments on model membranes. Commun Biol 5:1–13
120. Godino E, López JN, Zarguit I, Doerr A, Jimenez M, Rivas G, Danelon C (2020) Cell-free biogenesis of bacterial division proto-rings that can constrict liposomes. Commun Biol 3:1–11
121. Kattan J, Doerr A, Dogterom M, Danelon C (2021) Shaping liposomes by cell-free expressed bacterial microtubules. ACS Synth Biol 10:2447–2455
122. Dreher Y, Jahnke K, Bobkova E, Spatz JP, Göpfrich K (2021) Division and regrowth of phase-separated giant unilamellar vesicles**. Angew Chem 133:10756–10764
123. Steinkühler J, Knorr RL, Zhao Z, Bhatia T, Bartelt SM, Wegner S, Dimova R, Lipowsky R (2020) Controlled division of cell-sized vesicles by low densities of membrane-bound proteins. Nat Commun 11:905
124. Kurokawa C, Fujiwara K, Morita M, Kawamata I, Kawagishi Y, Sakai A, Murayama Y, Nomura SM, Murata S, Takinoue M et al (2017) DNA cytoskeleton for stabilizing artificial cells. Proc Natl Acad Sci 114:7228–7233
125. Jahnke K, Huth V, Mersdorf U, Liu N, Göpfrich K (2021) Bottom-up assembly of synthetic cells with a DNA cytoskeleton. ACS Nano. https://doi.org/10.1021/acsnano.1c10703
126. Litschel T, Kelley CF, Holz D, Adeli Koudehi M, Vogel SK, Burbaum L, Mizuno N, Vavylonis D, Schwille P (2021) Reconstitution of contractile actomyosin rings in vesicles. Nat Commun 12:2254
127. Elowitz MB, Levine AJ, Siggia ED, Swain PS (2002) Stochastic gene expression in a single cell. Science 297:1183–1186
128. Bar-Even A, Paulsson J, Maheshri N, Carmi M, O'Shea E, Pilpel Y, Barkai N (2006) Noise in protein expression scales with natural protein abundance. Nat Genet 38:636–643
129. Hansen MMK, Meijer LHH, Spruijt E, Maas RJM, Rosquelles MV, Groen J, Heus HA, Huck WTS (2015) Macromolecular crowding creates heterogeneous environments of gene expression in picolitre droplets. Nat Nanotechnol 11:191–197

130. Nishimura K, Tsuru S, Suzuki H, Yomo T (2015) Stochasticity in gene expression in a cell-sized compartment. ACS Synth Biol 4:566–576
131. Wang S, Majumder S, Emery NJ, Liu AP (2018) Simultaneous monitoring of transcription and translation in mammalian cell-free expression in bulk and in cell-sized droplets. Synth Biol 3: 1–9
132. Sakamoto R, Noireaux V, Maeda YT (2018) Anomalous scaling of gene expression in confined cell-free reactions. Sci Rep 8:7364
133. Gonzales DT, Yandrapalli N, Robinson T, Zechner C, T-YD T (2022) Cell-free gene expression dynamics in synthetic cell populations. ACS Synth Biol 11:205–215
134. Vibhute MA, Schaap MH, Maas RJM, Nelissen FHT, Spruijt E, Heus HA, Hansen MMK, Huck WTS (2020) Transcription and translation in cytomimetic protocells perform most efficiently at distinct macromolecular crowding conditions. ACS Synth Biol 9:2797–2807
135. Deng NN, Vibhute MA, Zheng L, Zhao H, Yelleswarapu M, Huck WTS (2018) Macromolecularly crowded protocells from reversibly shrinking monodisperse liposomes. J Am Chem Soc 140:7399–7402
136. Matthies D, Haberstock S, Joos F, Dötsch V, Vonck J, Bernhard F, Meier T (2011) Cell-free expression and assembly of ATP synthase. J Mol Biol 413:593–603
137. Berhanu S, Ueda T, Kuruma Y (2019) Artificial photosynthetic cell producing energy for protein synthesis. Nat Commun 10
138. Sakatani Y, Yomo T, Ichihashi N (2018) Self-replication of circular DNA by a self-encoded DNA polymerase through rolling-circle replication and recombination. Sci Rep 8:1–5
139. Okauchi H, Ichihashi N (2021) Continuous cell-free replication and evolution of artificial genomic DNA in a compartmentalized gene expression system. ACS Synth Biol 10:3507–3517
140. Van Nies P, Westerlaken I, Blanken D, Salas M, Mencía M, Danelon C (2018) Self-replication of DNA by its encoded proteins in liposome-based synthetic cells. Nat Commun 9:1–12
141. Libicher K, Hornberger R, Heymann M, Mutschler H (2020) In vitro self-replication and multicistronic expression of large synthetic genomes. Nat Commun 11:1–8
142. Fujii R, Kitaoka M, Hayashi K (2006) Error-prone rolling circle amplification: the simplest random mutagenesis protocol. Nat Protoc 1:2493–2497
143. Kita H, Matsuura T, Sunami T, Hosoda K, Ichihashi N, Tsukada K, Urabe I, Yomo T (2008) Replication of genetic information with self-encoded replicase in liposomes. Chembiochem 9: 2403–2410
144. Mizuuchi R, Ichihashi N (2020) Translation-coupled RNA replication and parasitic replicators in membrane-free compartments. Chem Commun 56:13453–13456
145. Caschera F, Karim AS, Gazzola G, D'Aquino AE, Packard NH, Jewett MC (2018) High-throughput optimization cycle of a cell-free ribosome assembly and protein synthesis system. ACS Synth Biol 7:2841–2853
146. Caschera F, Lee JW, Ho KKY, Liu AP, Jewett MC (2016) Cell-free compartmentalized protein synthesis inside double emulsion templated liposomes with in vitro synthesized and assembled ribosomes. Chem Commun 52:5467–5469
147. Niederholtmeyer H, Chaggan C, Devaraj NK (2018) Communication and quorum sensing in non-living mimics of eukaryotic cells. Nat Commun 9:5027
148. Dupin A, Simmel FC (2019) Signalling and differentiation in emulsion-based multi-compartmentalized in vitro gene circuits. Nat Chem 11:32–39
149. Dupin A, Aufinger L, Styazhkin I, Rothfischer F, Kaufmann BK, Schwarz S, Galensowske N, Clausen-Schaumann H, Simmel FC (2022) Synthetic cell–based materials extract positional information from morphogen gradients. Sci Adv 8:1–19
150. Jung K, Fabiani F, Hoyer E, Lassak J (2018) Bacterial transmembrane signalling systems and their engineering for biosensing. Open Biol 8
151. Peruzzi JA, Steinkühler J, Vu TQ, Gunnels TF, Lu P, Baker D, Kamat NP (2022) Hydrophobic mismatch drives self-organization of designer proteins into synthetic membranes. bioRxiv. https://doi.org/10.1101/2022.06.01.494374

152. Steinkühler J, Peruzzi JA, Krüger A, Jacobs ML, Michael C (2023) Improving cell-free expression of membrane proteins by tuning ribosome co-translational membrane association and nascent chain aggregation. bioRxiv. https://doi.org/10.1101/2023.02.10.527944
153. Peruzzi JA, Galvez NR, Kamat NP (2022) Engineering transmembrane signal transduction in synthetic membranes using two-component systems. bioRxiv
154. Jayaraman P, Yeoh JW, Jayaraman S, Teh AY, Zhang J, Poh CL (2018) Cell-free optogenetic gene expression system. ACS Synth Biol 7:986–994
155. Adir O, Albalak MR, Abel R, Weiss LE, Chen G, Gruber A, Staufer O, Kurman Y, Kaminer I, Shklover J et al (2022) Synthetic cells with self-activating optogenetic proteins communicate with natural cells. Nat Commun 13:2328
156. Krinsky N, Kaduri M, Zinger A, Shainsky-Roitman J, Goldfeder M, Benhar I, Hershkovitz D, Schroeder A (2018) Synthetic cells synthesize therapeutic proteins inside tumors. Adv Healthc Mater 7:1701163
157. Lussier F, Staufer O, Platzman I, Spatz JP (2021) Can bottom-up synthetic biology generate advanced drug-delivery systems? Trends Biotechnol 39:445–459

Adv Biochem Eng Biotechnol (2023) 186: 103–120
https://doi.org/10.1007/10_2023_228

Published online: 29 August 2023

Cell-Free Synthesis and Electrophysiological Analysis of Multipass Voltage-Gated Ion Channels Tethered in Microsomal Membranes

Yogesh Pandey, Srujan Kumar Dondapati, Doreen Wüstenhagen, and Stefan Kubick

Contents

1 Introduction 105
2 Materials 111
  2.1 CFPS Using CHO and *Sf*21 Lysate 111
  2.2 Quantitative Analysis of Synthesized Proteins 111
  2.3 Preparation of Proteoliposomes 112
  2.4 Single-Channel Analysis on Planar Lipid Bilayers 112
3 Method 112
  3.1 Cell-Free Synthesis 112

Y. Pandey
Fraunhofer Institute for Cell Therapy and Immunology (IZI), Branch Bioanalytics and Bioprocesses (IZI-BB), Potsdam, Germany

Institut für Biochemie und Biologie, University of Potsdam, Potsdam, OT Golm, Germany

S. K. Dondapati (✉) and D. Wüstenhagen
Fraunhofer Institute for Cell Therapy and Immunology (IZI), Branch Bioanalytics and Bioprocesses (IZI-BB), Potsdam, Germany
e-mail: srujan.dondapati@izi-bb.fraunhofer.de

S. Kubick
Fraunhofer Institute for Cell Therapy and Immunology (IZI), Branch Bioanalytics and Bioprocesses (IZI-BB), Potsdam, Germany

Institute of Biotechnology, Technische Universität Berlin, Berlin, Germany

Institute of Chemistry and Biochemistry-Biochemistry, Freie Universität Berlin, Berlin, Germany

Faculty of Health Science, Joint Faculty of the Brandenburg University of Technology Cottbus-Senftenberg, The Brandenburg Medical School Theodor Fontane and the University of Potsdam, Potsdam, Germany

3.2 Quantification and Qualitative Analysis of the de novo Synthesized Protein 113
3.3 Preparation of Proteoliposomes 114
3.4 Formation of Lipid Bilayers 114
3.5 Electrophysiological Activity Measurement and Functional Analysis 115
4 Notes 115
5 Discussion and Future Implications of the Eukaryotic CFPS System 117
References 118

Abstract Cell-free protein synthesis (CFPS) has emerged as a powerful tool for the rapid synthesis and analysis of various structurally and functionally distinct proteins. These include 'difficult-to-express' membrane proteins such as large multipass ion channel receptors. Owing to their membrane localization, eukaryotic CFPS supplemented with endoplasmic reticulum (ER)-derived microsomal vesicles has proven to be an efficient system for the synthesis of functional membrane proteins. Here we demonstrate the applicability of the eukaryotic cell-free systems based on lysates from the mammalian Chinese Hamster Ovary (CHO) and insect *Spodoptera frugiperda* (*Sf*21) cells. We demonstrate the efficiency of the systems in the *de novo* cell-free synthesis of the human cardiac ion channels: ether-a-go-go potassium channel (hERG) $K_V11.1$ and the voltage-gated sodium channel $hNa_V1.5$.

Graphical Abstract

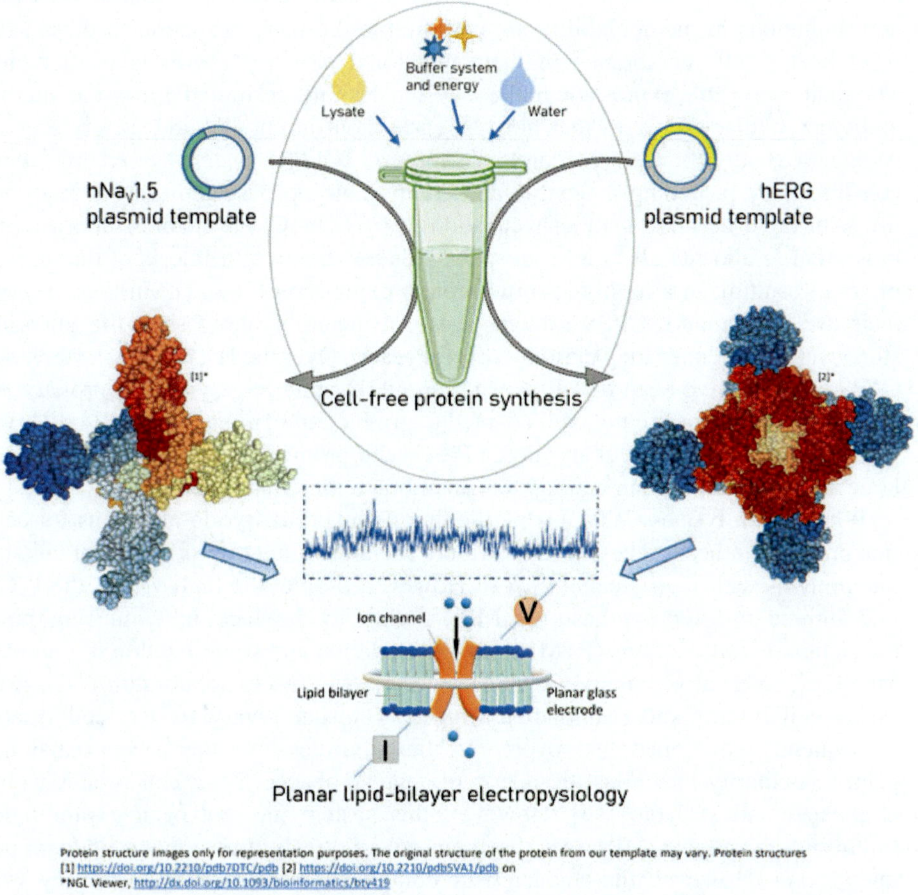

Keyword Cell-free protein synthesis, Eukaryotic lysates, Ion channels, Membrane proteins, Microsomes, Planar lipid bilayer

1 Introduction

Membrane proteins (MPs) account for about 30% of the human proteome owing to diverse and crucial roles in vital cellular processes [1–3]. Implications of MPs in diseases make them high-value druggable targets with the majority of global therapeutics targeted at MPs [4]. However, the detailed structural and functional information of MPs remains unexplored largely due to difficulties in synthesizing and native purification of high-quality MPs [5–7]. The cell-based expression has been the conventional method for the synthesis of proteins. However, factors such as extensive culturing facilities, specialized media requirements, time-to-expression, and limitations in manipulability are undesirable. Further, extraction of these MPs from hosts' cell membranes in their functional structural form is challenging. Maximal achievable expression in these systems is further limited due to the internal and vital cellular processes that must be carried out in parallel and factors such as cellular toxicity [8]. Cell-free protein synthesis (CFPS) systems overcome these constraints by providing a flexible and controllable open system that is ready-to-use with rapid de novo synthesis capabilities [9]. The CFPS enables the synthetic biosystem to allocate all of its energy towards the de novo synthesis of the desired protein, resulting in a highly specific protein expression [10, 11]. Further, beyond their use on demand, CFPS systems have an incredibly long shelf-life allowing storage at −80°C over long periods (several years). Over the last decade, a variety of CFPS systems have been developed and tested. Despite large yields in prokaryotic CFPS, these systems are not well adapted to production of eukaryotic MPs [12]. We will hence focus on the eukaryotic CFPS for the production of the two exemplary large and complex human cardiac transmembrane ion-channel proteins.

Wheat germ lysates (WGL) were the first popular eukaryotic platforms for cell-free protein synthesis. They became widely acclaimed due to production of eukaryotic proteins with high yields [13, 14]. However, despite the high yields, the WGL find limited use for synthesis of MPs, owing to the lack in conferring post-translational modifications (PTMs) like glycosylation and solubilization of complex MPs [12]. Alternative eukaryotic CFPS systems, viz. Insect *Spodoptera frugiperda (Sf*21) cell lysates and mammalian Chinese Hamster Ovary (CHO) cell lysates subsequently developed that overcome these hurdles. Another system that has gained popularity in the last decade is the plant Tobacco BY-2 cell lysates which is cheaper and scalable [15]. However, this system has not been explored for synthesis of complex MPs. All the three eukaryotic lysates contain endogenous microsomes. However, the characteristic compositions of these systems may vary due to the different sources of origin. *Sf*21 lysates contain translationally active endogenous ER membranes that support the signal peptide-mediated translocation of proteins across the membrane, and further provide functions such as signal peptide cleavage, PTMs like N-glycosylation, and lipid modifications [16–18]. CHO cell-based expression is well established and approved for the large-scale synthesis of several biologics by the FDA because it undergoes human-compatible PTMs [19–21]. CFPS systems based on CHO lysates have thus evolved as the go to systems for the expression of difficult-to-express proteins especially when the more complex PTMs are desirable. CHO-based CF systems retain most of the features of CHO cells while adding flexibility due to the lack of redundant components such as cell membrane boundaries. CHO-based lysates much like their insect counterparts harbour endogenous microsomal vesicles enabling translocation of transmembrane proteins and secretory proteins. Thus, using CHO cell lysate for CFPS has a potential value and enables new opportunities, in particular the high-yield production of pharmaceutically relevant MPs. There is a significant increase in the number of publications based on *Sf*21 and CHO lysates for CFPS [17, 22–27].

In this chapter, we present methods for de novo synthesis and the functional assay of two critical cardiac ion channels: the human voltage-gated sodium channel 1.5 (hNa$_V$1.5) and the human ether-a-go-go channels, hERG. We studied these large MPs via production in insect-based *Sf*21 and CHO-based cell-free systems. The expression and analysis of the two large and complex human cardiac ion channels are depicted as an exemplary application of the two CFPS systems.

Cardiac channelopathies are abundant owing to a large repertoire of ion-channels and phenotypes that attribute to a dysfunctional heart. Cardiac ion-channels are expressed in cardiomyocytes which predominantly occupy up to 85% of heart's volume [28]. Cardiac excitation–contraction coupling in cardiomyocytes is a complex process involving a diverse set of ion-channels which conduct a variety of ions, viz. Na +, Ca 2+, K+ [29]. Amongst these, the rapid inward sodium current (I_{Na}) conducted via the cardiac voltage-gated sodium channel Na$_v$1.5 (encoded by *SCN5A*) is pivotal for initiation of the excitation–contraction coupling. Na$_V$1.5 is one of the nine voltage-gated sodium channel-alpha subunits (VGSC-α family). Over 400 genetic variants in *SCN5A* attribute to cardiac dysfunctions resulting in life-threatening arrhythmias and structural heart disease [30–34]. Dysfunction of Na$_V$1.5 accounts for a variety of non-cardiac diseases, including ataxia, epilepsy, pain, bowel syndrome and myotonic dystrophy, and more recently cancer has further been associated with a dysfunctional hNa$_V$1.5 [35].

Amongst the most commonly assayed cardiac channels is the Human ether a go-go-related gene (hERG; encoded by *KCNH2*) K+ channels. hERG conducts the rapid delayed rectifier potassium current and mediates action potential repolarization in the heart. hERG contributes to the second most prevalent form of congenital long QT syndrome (LQTS), LQTS2 [36]. Aside from cardiac muscle cells, voltage-gated *hERG*-encoded Kv11.1 potassium channels are expressed in several tissues throughout the body including endocrine, pancreatic cells, and cancerous cells [37–39]. hERG channel has a high ligand promiscuity and hence, any new drug that is designed must be assayed for interaction with hERG. A drug-induced blockage of hERG can cause a ventricular tachycardia torsade de pointes (TdP) [40]. hERG assay has hence emerged as a critical step in drug development to avoid cardiac- and hepato-toxicity [41]. Comprehensive in vitro proarrhythmia assays (CiPA) have largely depended on assaying cardiac therapeutics against hERG toxicity [42].

The proteins synthesized here are both 'difficult to express' due to their large size and complex conformation. Both the protein assemblies differ with some shared similarities in the functional channel structure. The hERG forms a 135–156 kDa, 6 transmembrane protein subunit which assembles as a tetramer to form the functional potassium channel (Fig. 1). The hNa$_V$1.5 is a 220 kDa, 24 transmembrane protein comprising of 4 domains with 6 segments each (Fig. 2). Despite the large sizes, we demonstrate that endogenous ER-derived microsomes greatly support the

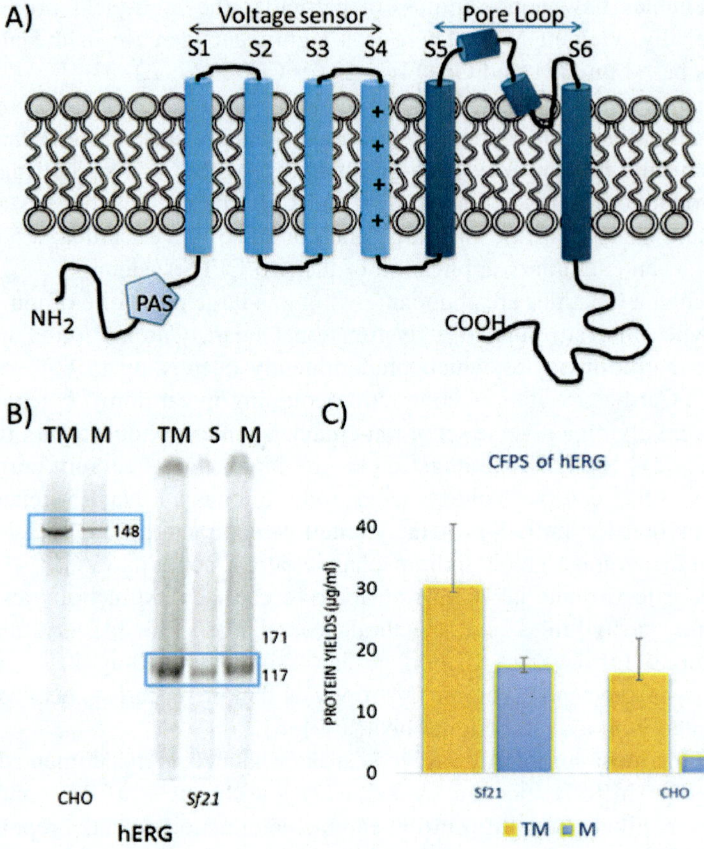

Fig. 1 (**a**) Canonical structure of the hERG voltage-gated potassium channel depicting 6 trans-membrane segments S1–S6; S1–S4 function as the voltage-sensing domain while the S5–S6 form the pore loop. (**b**) Autoradiograph of hERG channel protein synthesized with CHO CFPS and *Sf*21 CFPS as visualized using a phosphorimager on 10% and 4–16% SDS-PAGE gels. Samples were loaded as either the **TM**-Translational reaction mix, the **S**-Soluble fraction which was collected as a supernatant after a high-speed centrifugation of the sample, and **M**-Microsomal fraction which was collected as a pellet after the centrifugation and resuspended in PBS. Please note that only TM and M were run for the CHO CFPS reaction for the hERG protein. Differences in the observed protein size may be attributed to differences in mobility in a uniform vs. gradient gel and the glycosylation status of the protein in the two different lysates. 150 kDa band depicts a mature completely glycosylated membrane localized hERG, whereas a 135 kDa band may depict a core glycosylated band. (**c**) Protein yields as measured via scintillation counting of the incorporated ^{14}C Leucine

integration and assembly of the MPs ensuring that they are hosted in a close to the native environment [43]. Further electrophysiological analysis revealed that the protein-enriched microsomes present in the *Sf*21 and CHO lysates can be directly tested on artificial planar lipid bilayers or as proteoliposome assemblies. These ion

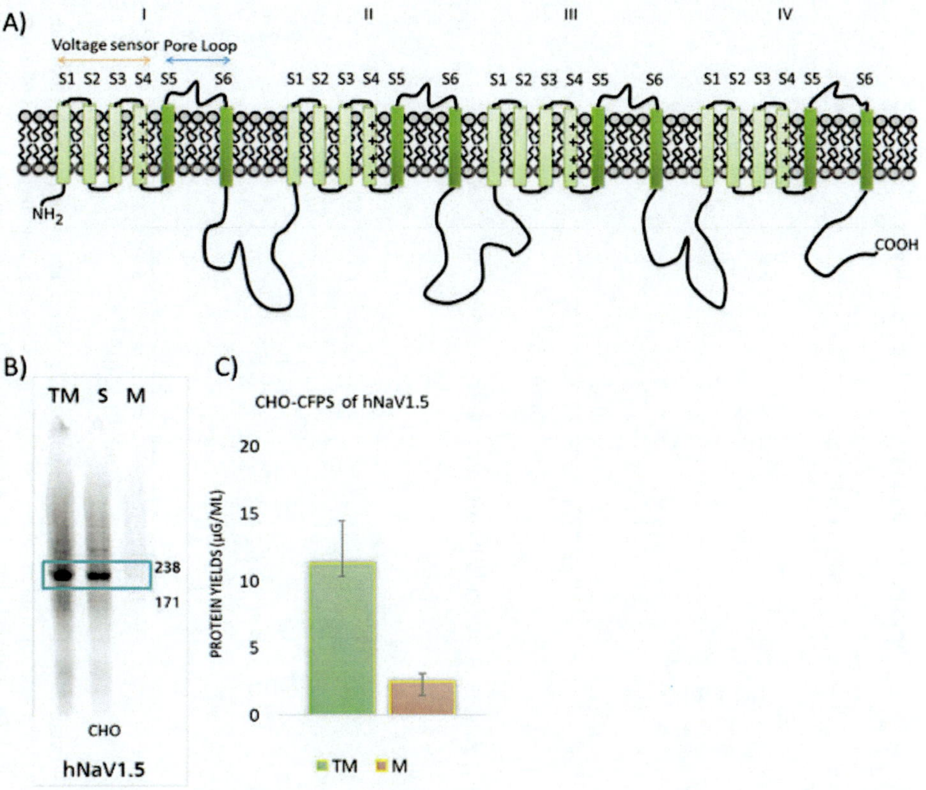

Fig. 2 (**a**) Canonical structure of the 24 transmembrane $hNa_V1.5$ voltage-gated sodium channel comprising of 4 repetitive domains I–IV with 6 segments S1–S6 each. The four domains assemble together with the S1–S4 regions forming the voltage-sensing domain and the S5–S6 segments conforming to the central pore domain of the channel. (**b**) $hNa_V1.5$ visualized on a 4–16% SDS-PAGE gel, and (**c**) Protein yields from CHO-CFPS of $hNa_V1.5$ (figure adapted from Pandey, Y et al. 2023)

channels exhibited native properties of the respective protein which could be validated with channel-specific blockers (Fig. 3). Owing to a rapid synthesis, co-translational translocation into native membranes, and ease of analysis with electrophysiology, we suggest the further implication of the CHO and *Sf*21 lysate systems for the synthesis of other large multipass voltage-gated ion channels.

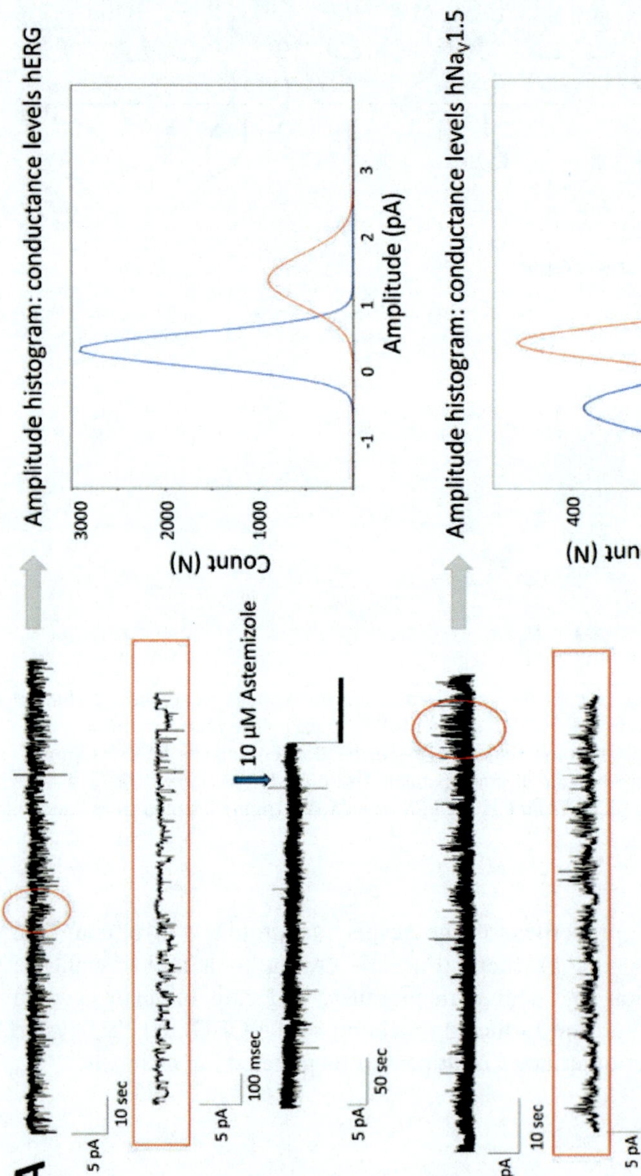

Fig. 3 Single channel activity on DPhPC bilayers with the inset (red) showing respective zoomed in regions of the traces, and corresponding amplitude histograms (panels on the right) for (**a**) DOTAP Proteoliposomes reconstituted from enriched microsomes containing the de novo synthesized hERG potassium channel using the *Sf*21 CFPS in response to a voltage pulse of +100 mV, and (**b**) Crude enriched microsomes from CHO-CFPS of voltage gated-potassium channel hNa$_V$1.5 in response to a voltage pulse of +40 mV. Bottom panels depict blockage of the protein elicited activity with the addition of 10 μM Astemizole and 5 mM Lidocaine for the hERG and hNa$_V$1.5 channels, respectively

2 Materials

2.1 *CFPS Using CHO and Sf21 Lysate*

1. Ice container
2. 2.5 ml reaction tubes
3. *Sf*21 and CHO lysate prepared as described previously (see Note 1, 2). Flash-freeze in liquid nitrogen after each use. Store at −80°C (See Note 3)
4. T7 RNA polymerase
5. 10× translation mix: 300 mM HEPES-KOH (pH 7.6), 390 mM $Mg(OAc)_2$ (See Note 4), 1800 mM KOAc, 2.5 mM spermidine, 1 mM of each canonical amino acid. Store at −80°C
6. 5× energy mix: 0.5 mg/ml creatine phosphokinase, 100 mM creatine phosphate, 1.5 mM GTP,1.5 mM CTP, 1.5 mM UTP, 8.75 mM ATP, and 0.5 mM m7G (*ppp*)G cap analog. Store at −80°C
7. ^{14}C-leucine (200 dpm/pmol, specific radioactivity of 46.15 for both CHO and *Sf*21 CFPS). Store at −20°C
8. Plasmid encoding K_v11.1 (hERG) and SCN5A (hNa$_V$1.5) (see Note 5–6). Store at −20°C
9. Ultrapure Milli-Q water
10. Thermomixer incubator
11. Phosphate-buffered saline (PBS). Store at 4°C

2.2 *Quantitative Analysis of Synthesized Proteins*

1. Trichloroacetic acid (TCA)
2. Casein hydroxylase
3. Filter paper (MN GF-3, Macherey-Nagel)
4. Acetone
5. Water bath
6. Glass test tubes
7. Vacuum filtration system
8. Scintillation tubes (Zinsser Analytic)
9. Scintillation cocktail (Quicksafe A, Zinsser Analytic)
10. LS6500 Multi-Purpose scintillation counter (Beckman Coulter)
11. Orbital shaker
12. NuPAGE® LDS Sample Buffer (Invitrogen)
13. Precast SDS-PAGE gels (NuPAGE 4–16% Bis–Tris Gel with MES SDS buffer, Invitrogen)
14. Self-cast 10% SDS-PAGE gel
15. Fluorescently labelled protein ladder for SDS-PAGE
16. SDS-PAGE running buffer: 50 mM MES, 50 mM Tris Base, 0.1% SDS, 1 mM EDTA, pH 7.3

17. SDS-PAGE gel tank system
18. Gel dryer
19. Radioactive ink
20. Phosphor screens
21. Fluorescence/phosphorimager (Typhoon TRIO + Imager, GE Healthcare)

2.3 Preparation of Proteoliposomes

1. Dioleoyl-3-trimethylammonium propane (DOTAP) lipids (15 mM) in Cholate buffer (100 mM sodium cholate, 20 mM Tris, 100 mM NaCl, pH 7.4)
2. Biobeads SM-2 (Bio-Rad)

2.4 Single-Channel Analysis on Planar Lipid Bilayers

1. Orbit 16 System (Nanion Technologies GmbH, Munich, Germany)
2. 1,2-diphytanoyl-sn-glycero-3-phosphocholine(DPhPC) lipids at a concentration of 10 mg/ml dissolved in Octane (SigmaAldrich)
3. Recording solutions 150 mM KCl or 150 mM NaCl in Phosphate-buffered saline at pH 7.4
4. Multi electrode cavity arrays (MECA) chips (Nanion GmbH, Germany)
5. Amplifier systems: EPC-10 amplifier, (HEKA Electronic Dr. Schulze GmbH, Lambrecht, Germany) and the data acquisition software Patchmaster (HEKA), or Elements Multichannel amplifier and data acquisition software Elements Data Reader (Elements SRL, Italy)
6. Clampfit 11.0.3 software (Molecular devices, Sunnyvale, California, USA) for electrophysiology analysis

3 Method

3.1 Cell-Free Synthesis

Coupled transcription–translation 'one-pot' format (batch-based) cell-free synthesis reactions were assembled as described below.

3.1.1 Batch-Based CFPS with *Sf*21 Lysate

1. Thaw all components of the cell-free reaction on ice. Mix all components thoroughly before usage

2. Pipet the following components on ice using a 1.5 ml reaction vessel to set up a reaction: PolyG (f.c. 20 μM), 5 μl 10× translation mix (f.c. 1×), *Sf*21 lysate (f.c. 40%), T7 RNA polymerase (f.c. 1 U/μl), and 5x energy mix (f.c. 1×). Mix the components in the vial thoroughly after addition of each component
3. Add 200 dpm/pmol ^{14}C-leucine (f.c. 30 μM) for subsequent qualitative analysis by autoradiography
4. Adjust the final volume of the reaction mix with ultrapure water to the desired volume
5. Mix all components thoroughly and incubate the reaction at 27°C and 500 rpm for 2 h in a thermomixer. Post reaction incubation, the reactions are placed on ice for further processing

3.1.2 Batch-Based CFPS with CHO Lysate

Setting up of the CHO CFPS is similar to described for the *Sf*21 CFPS reaction setup. However, the incubation conditions of the reaction differ (see Note 7). CHO CFPS is generally carried out at 27°C or 30°C and 600 rpm for 3 h in a thermomixer.

3.2 *Quantification and Qualitative Analysis of the de novo Synthesized Protein*

All the quantitative and qualitative measurements were performed by TCA precipitation of ^{14}C-leucine incorporated proteins and SDS-PAGE combined with autoradiography.

1. 2× 5 μl of the reaction mixture (TM) is collected for TCA precipitation and 1× 5 μl was collected for subsequent SDS-PAGE analysis
2. The remaining reaction mixture is centrifuged at 16,000 × *g* for 10 min at 4°C
3. The supernatant (S) is transferred into a separate Eppendorf tube
4. 2× 5 μl of the supernatant is collected for TCA precipitation and 1× 5 μl is collected for subsequent SDS-PAGE analysis
5. The pelleted vesicular fraction (M) that remains in the Eppendorf tube is suspended to the original volume with Phosphate-Buffered Saline (PBS)
6. 2× 5 μl of the suspended M fraction is collected for TCA precipitation and 1× 5 μl is collected for subsequent SDS-PAGE analysis
7. For TCA precipitation, the aliquots from the three fractions (TM, S, and M) are incubated in 3 ml trichloroacetic acid (TCA) each in a water bath at 80°C for 15 min. Thereafter, the TCA containing aliquots are cooled on ice for 30 min
8. A vacuum filtration system is used to collect the radiolabelled TCA precipitated proteins on filter paper. The tubes are rinsed twice with TCA to collect any remnants of the protein
9. Acetone is used in a final step to wash any further remnants of TCA and to aid drying of the filter papers under the vacuum manifold

10. The dried protein retaining filter papers are transferred to scintillation tubes (Zinsser Analytic) and 3 ml of scintillation cocktail is added
11. The tubes are shaken on an orbital shaker for at least 1 h
12. The quantification of incorporation of ^{14}C-leucine is carried out by liquid scintillation counting using the scintillation counter
13. For SDS-PAGE analysis, the collected fractions are precipitated using cold acetone: 45 μl of the water along with 150 μl of ice cold acetone are added to each of the 5 μl aliquots and incubated on ice for 15 min. The tubes are then centrifuged at 16,000 × *g* for 10 min at 4°C. The supernatant containing the acetone is discarded retaining the precipitated protein pellets
14. The pellets are resuspended in 20 μl of NuPAGE® LDS Sample Buffer and loaded onto precast SDS-PAGE gels. Run the gel at 200 V for 35 min or till the running front has visually reached 90% of the gel length. Dry the gels for 70 min at 70°C using the Unigeldryer
15. The radioactively labelled proteins can be visualized directly in-gel using the phosphorimager or visualized post incubation for 24 h by placing them on a phosphoscreen

3.3 Preparation of Proteoliposomes

1. 25 μl of hERG incorporated microsomes were mixed with 25 μl of the DOTAP lipids (suspended to 15 mM sodium cholate) and mixed at RT for 30 min
2. Biobeads-SM2 were added to the mix up to 80% volume and incubated for 45 min at RT
3. The supernatant was used for functionality measurements on lipid bilayers. Control samples were CFPS reactions without any DNA template resulting in with null protein yields

3.4 Formation of Lipid Bilayers

1. DPhPC lipids are dissolved in octane at a concentration of 10 mg/ml (see Note 8). All the stocks of lipids are stored at −20°C
2. Lipid bilayers are formed on MECA array chips mounted on the Orbit 16 System
3. Lipid bilayers are formed: 200 μl of electrolyte solution is added to the measurement chamber containing the MECA chip. Once the buffer is added, all the electrodes will be in open (seal resistance of few MΩ)
4. For the automated bilayer formation on the 16 cavities in parallel, a small amount of approx. 0.1 μl of DPhPC at 10 mg/ml in octane is pipetted on to the chip surface and painted with the help of bubble method (see Note 9)
5. Lipid bilayer formation will be indicated by the change in resistance from MΩ to GΩ (see Notes 10)

3.5 *Electrophysiological Activity Measurement and Functional Analysis*

1. Once the lipid bilayer is formed, add 1–5 μl of the enriched native microsomes or the DOTAP proteoliposomes are applied directly into the buffer chamber containing the lipid bilayers and wait for fusion.
2. After incubating the lipid bilayer with the cell-free produced sample, measure activity from the voltage-clamped lipid bilayers of the de novo synthesized protein by a single channel amplifier single channel amplifier (EPC-10) or the multiplex amplifier from Elements SRL. The data acquisition is done using the software Patchmaster or Elements Data reader, respectively. Recordings are done at a sampling rate of 5–50 kHz with a 10 kHz Bessel filter or a 500 Hz virtual visual filter. Single channel currents are recorded from the voltage-clamped lipid bilayers (see Note 11)
3. Data analysis: Voltage potential protocols were applied for monitoring currents recorded using Patchmaster software and analysed by the Clampfit software version 11.0.3.

4 Notes

1. The method of preparation for CHO lysates has previously been described [44]. In summary, CHO cells are grown at 37°C in a serum-free cell medium using a Biostat BDCU II bioreactor (Sartorius Stedim Biotech GmbH). Cells are harvested when the culture reached a density of 3.5–5 × 10^6 cells/ml by centrifugation at 200 × *g* for 5 min. Cell pellet was washed twice and resuspended in a homogenization buffer (40 mM HEPES-KOH (pH 7.5), 100 mM NaOAc, and 4 mM DTT). Resuspended CHO cells were lysed mechanically by passing them through a 20-gauge needle using a syringe. Post lysis, the homogenate was centrifuged at 6500 × *g* for 10 min to remove the cell nuclei and cellular debris. The resulting supernatant was applied to an equilibrated Sephadex G-25 column (GE Healthcare, Freiburg, Germany). Eluted fractions containing the highest RNA/protein ratios were pooled. Endogenous mRNA was removed by treating the cell lysates with S7 micrococcal nuclease (f.c. 10 U/ml) and $CaCl_2$ (f.c. 1 mM) for 2 min at RT. The micrococcal nuclease was deactivated with the addition of EGTA (f.c. 6.7 mM). The lysate was supplemented with creatine kinase (f.c. 100 μg/ml) as an energy regeneration component. Aliquots of the CHO lysate were flash-freezed in liquid nitrogen and stored at −80°C.
2. Preparation of *Sf*21 lysate [45–47]: Fall armyworm (*Spodoptera frugiperda* 21, *Sf*21) cells were grown in an animal component free insect cell medium (Insect-XPRESS Medium, Lonza) in a controlled fermenter at 27°C. Cells were harvested at a density of approximately 4 × 10^6 cells/ml. Sf21 cells were

collected by centrifugation at 200 × *g* for 10 min, washed and resuspended with a homogenization buffer containing DTT (40 mM HEPES–KOH (pH 7.5), 100 mM KOAc, 4 mM DTT. *Sf*21 cells were disrupted mechanically by passing suspension through a 20-gauge needle followed by centrifugation at 10,000 × *g* for 10 min to remove the nuclei and cell debris. The resulting supernatant was applied to an equilibrated Sephadex G-25 column (GE Healthcare, Freiburg, Germany). Fractions with the highest RNA content were pooled. Cell lysates were treated with micrococcal nuclease (S7) in order to remove residual endogenous mRNA. Finally, aliquots of the cell lysate were immediately flash-freezed in liquid nitrogen and stored at −80°C.

3. The lysates must be stored at −80°C as aliquots in smaller volumes to avoid repeated freeze–thaw cycles that may be detrimental to the quality of the lysate.
4. The concentrations of the translational mix may be optimized especially when trying a new batch of lysates. For instance, we tested and used 330 mM (10×) and 495 mM (15×) of $Mg(OAc)_2$.
5. The DNA template used for CFPS requires a T7 promotor and terminator. The final concentration of the template in the reaction influences protein synthesis efficiency and hence, this may need optimization according to the protein of interest.
6. The *Sf*21 based cell-free synthesis does not necessarily require an IRES site. However, CHO synthesis greatly benefits from inclusion of an Internal ribosome entry site of Cricket Paralysis virus (IRES-CRPV) site [21]. This must be considered when designing the CDS for expression in CHO CFPS systems.
7. In general, the optimal temperature of the CHO cell-free system is 30°C and 27°C for *Sf*21 lysates [44] but the optimal incubation temperature may be different for individual proteins, lysate batches, and the buffer system.
8. The choice of lipids used for the electrophysiological assay on artificial bilayers depends on the protein of interest. Proteins such as voltage-gated ion channels are sensitive to the lipid composition [48, 49] and different lipids may be tried to improve the activity of the protein during electrophysiological analysis.
9. The bubble method has been described previously [50]. Either the painting method or the bubble method was used for building the bilayers. For this 0.1 μl of lipid is taken in a pipette and a bubble is formed and retracted over the electrode cavity.
10. To avoid lipid-choked bilayers, post formation of bilayers a ZAP pulse (+200 mV for 200 ms) may be used to break the bilayers and repaint stable bilayers. If there is no effect of the ZAP pulse, it indicates the clogging of the electrodes by the solvent or formation of multiple layers of lipids. Further, empty bilayers can be tested for artefactual activity by testing them with voltage pulses of +100 and −100 mV to rule out any voltage-dependent activity emanating from the lipid bilayers.
11. When recording from multiple bilayers at the same time, it is recommended to shut off the voltage command on over-active bilayers or to shut the bilayer completely to avoid other bilayers from breaking due to the cross-talk between the bilayers.

5 Discussion and Future Implications of the Eukaryotic CFPS System

In the last decade, a variety of cell-free platforms derived from eukaryotic organisms have been developed and tested. Some of these, like the wheat germ lysate system and the more recently developed Tobacco BY-2 cell lysates, have demonstrated the use and scalability of CFPS for de novo protein synthesis. However, only the *Sf*21 and CHO lysate systems have been the system of choice and extensively used to synthesize complex MPs such as ion channels. Conventionally, the performance metrics of CFPS systems are correlated to the expression of a fluorescent marker protein to quantify the protein yields. Fluorescent marker proteins such as sf-GFP or eYFP have found use as fusion proteins owing to a remarkable quality of stable folding and assembly. However, in contrast, MPs are usually large and complex proteins that may require a complex network of accessory proteins and a membrane environment for their functional assembly. When it comes to MPs, the CHO and *Sf*21 CFPS platforms have proven to be the most tested and versatile cell-free systems available in the market today. Both systems offer high specificity for the production of eukaryotic proteins. The CHO lysate system is specifically specialized to enable 'human-like' post-translational modifications compared to any other eukaryotic cell-free system discussed here. CHO cells also enable closer to native-like glycosylation, are easy to transfect, and yield higher titres compared to the human embryonic kidney cells that largely facilitated their adoption [51]. A close competitor to these platforms is the tobacco-lysate platform which may be suitable for the expression of MPs due to the availability of microsomes and eukaryotic post-translational modifications. This system is also cheaper to produce than CHO lysates largely due to reduced maintenance and media costs. However, there are no reports on the synthesis of ion channels in this CFPS system thus far. Moreover, when human-like post-translational modifications are required, the CHO system due to its mammalian origin may be the best system to choose. Further studies and comparisons between the systems can elucidate specific capabilities and limitations of the systems concerning the proteins of interest being expressed. We recently also compared the functionality of protein-enriched endogenous microsomal vesicles present in the CHO cell-free lysate with protein-enriched native membrane vesicles prepared from CHO cells [22]. The results demonstrate the efficiency of the CHO-CFPS for synthesis and testing of 'difficult-to-express' ion channels and provide an insight into the similarities of the de novo synthesized protein to the natively expressed protein in CHO cells. *Sf*21 and CHO lysate systems have also recently demonstrated their efficiency in the synthesis of large-heteromeric channels [26, 27]. The studies so far elucidate the application of the CFPS platform towards highly complex MPs.

In the future, further studies need to be conducted to optimize these systems for large-scale production of therapeutically relevant MPs such as ion channels and transporters. With the recent spotlight on artificial intelligence-supported protein structure determination and design strategies, it may also soon be a regular practice to produce completely synthetic proteins [52]. Cell-free protein synthesis may not only provide a platform for the synthesis of complex synthetic proteins without the constraints of cellular viability and toxicity but also bring this technology to the desktop much like computers. The biggest hurdle that lies in the way is the adoption of the technology as in many cases, the CFPS platform may not be a comparable replacement to cellular systems unless scaled up. Costs associated with the technology hence need to be optimized to realize the commercial value of the system [53]. There are also efforts towards the development of novel lysate systems such as fungal lysates that may reduce the associated costs for production but due to their prokaryotic nature may be challenging to use when it comes to production of complex MPs. Hence, there is a need for finding the correct mix and match of systems and the protein of interest. There is a lot of scope to explore further in the future as cell-free technologies continue to gain interest.

Acknowledgements We would like to thank Wenzel, D for providing the lysates used. This research was funded by the Ministry of Science, Research and Culture (MWFK, Brandenburg, Germany), project PZ-Syn (project number F241-03-FhG/005/001), and the Federal Ministry of Education and Research (BMBF, Germany No. 031B0078A and 13GW0408C).

References

1. Gulezian E, Crivello C, Bednenko J et al (2021) Membrane protein production and formulation for drug discovery. Trends Pharmacol Sci 42:657–674
2. Marinko JT, Huang H, Penn WD et al (2019) Folding and misfolding of human membrane proteins in health and disease: from single molecules to cellular proteostasis. Chem Rev 119: 5537–5606
3. Almén MS, Nordström KJV, Fredriksson R et al (2009) Mapping the human membrane proteome: a majority of the human membrane proteins can be classified according to function and evolutionary origin. BMC Biol 7:50
4. Yin H, Flynn AD (2016) Drugging membrane protein interactions. Annu Rev Biomed Eng 18: 51–76
5. Kermani AA (2021) A guide to membrane protein X-ray crystallography. FEBS J 288:5788–5804
6. Yao X, Fan X, Yan N (2020) Cryo-EM analysis of a membrane protein embedded in the liposome. Proc Natl Acad Sci U S A 117:18497–18503
7. Werten PJL, Rémigy HW, de Groot BL et al (2002) Progress in the analysis of membrane protein structure and function. FEBS Lett 529:65–72
8. Eguchi Y, Makanae K, Hasunuma T et al (2018) Estimating the protein burden limit of yeast cells by measuring the expression limits of glycolytic proteins. eLife 7
9. Gregorio NE, Levine MZ, Oza JP (2019) A user's guide to cell-free protein synthesis. Methods Protoc 2
10. Carlson ED, Gan R, Hodgman CE et al (2012) Cell-free protein synthesis: applications come of age. Biotechnol Adv 30:1185–1194
11. Focke PJ, Hein C, Hoffmann B et al (2016) Combining in vitro folding with cell free protein synthesis for membrane protein expression. Biochemistry 55:4212–4219
12. Zemella A, Thoring L, Hoffmeister C et al (2015) Cell-free protein synthesis: pros and cons of prokaryotic and eukaryotic systems. Chembiochem 16:2420–2431

13. Harbers M (2014) Wheat germ systems for cell-free protein expression. FEBS Lett 588:2762–2773
14. Fogeron M-L, Lecoq L, Cole L et al (2021) Easy synthesis of complex biomolecular assemblies: wheat germ cell-free protein expression in structural biology. Front Mol Biosci 8:639587
15. Buntru M, Vogel S, Spiegel H et al (2014) Tobacco BY-2 cell-free lysate: an alternative and highly-productive plant-based in vitro translation system. BMC Biotechnol 14:37
16. Quast RB, Sonnabend A, Stech M et al (2016) High-yield cell-free synthesis of human EGFR by IRES-mediated protein translation in a continuous exchange cell-free reaction format. Sci Rep 6:30399
17. Dondapati SK, Stech M, Zemella A et al (2020) Cell-free protein synthesis: a promising option for future drug development. BioDrugs 34:327–348
18. Pool M, Dobberstein B (2001) Cotranslational translocation of proteins into microsomes: methods. In: John Wiley & Sons L (ed) eLS. Wiley
19. Tihanyi B, Nyitray L (2020) Recent advances in CHO cell line development for recombinant protein production. Drug Discov Today Technol 38:25–34
20. Zhu J (2012) Mammalian cell protein expression for biopharmaceutical production. Biotechnol Adv 30:1158–1170
21. Thoring L, Dondapati SK, Stech M et al (2017) High-yield production of "difficult-to-express" proteins in a continuous exchange cell-free system based on CHO cell lysates. Sci Rep 7:11710
22. Pandey Y, Dondapati SK, Kubick S (2023) Enriched cell-free and cell-based native membrane derived vesicles (nMV) enabling rapid in-vitro electrophysiological analysis of the voltage-gated sodium channel 1.5. Biochim Biophys Acta Biomembr 1865:184144
23. Dondapati SK, Pietruschka G, Thoring L et al (2019) Cell-free synthesis of human toll-like receptor 9 (TLR9): optimization of synthesis conditions and functional analysis. PloS One 14: e0215897
24. Dondapati SK, Lübberding H, Zemella A et al (2019) Functional reconstitution of membrane proteins derived from eukaryotic cell-free systems. Front Pharmacol 10:917
25. Haueis L, Stech M, Schneider E et al (2023) Rapid one-step capturing of native, cell-free synthesized and membrane-embedded GLP-1R. Int J Mol Sci 24
26. Ullrich J, Göhmann PJ, Zemella A et al (2022) Oligomerization of the heteromeric γ-aminobutyric acid receptor GABAB in a eukaryotic cell-free system. Sci Rep 12:20742
27. Ullrich J, Ohlhoff C, Dondapati S et al (2023) Evaluation of the ion channel assembly in a eukaryotic cell-free system focusing on two-pore domain potassium channels K2P. Int J Mol Sci 24:6299
28. Zhou P, Pu WT (2016) Recounting cardiac cellular composition. Circ Res 118:368–370
29. Eisner DA, Caldwell JL, Kistamás K et al (2017) Calcium and excitation-contraction coupling in the heart. Circ Res 121:181–195
30. Mikhailova VB, Karpushev AV, Vavilova VD et al (2022) Functional analysis of SCN5A genetic variants associated with Brugada syndrome. Cardiology 147:35–46
31. Li W, Yin L, Shen C et al (2018) SCN5A variants: association with cardiac disorders. Front Physiol 9:1372
32. Fernández-Falgueras A, Sarquella-Brugada G, Brugada J et al (2017) Cardiac channelopathies and sudden death: recent clinical and genetic advances. Biology (Basel) 6
33. Napolitano C, Bloise R, Monteforte N et al (2012) Sudden cardiac death and genetic ion channelopathies: long QT, Brugada, short QT, catecholaminergic polymorphic ventricular tachycardia, and idiopathic ventricular fibrillation. Circulation 125:2027–2034
34. Zaklyazminskaya E, Dzemeshkevich S (2016) The role of mutations in the SCN5A gene in cardiomyopathies. Biochim Biophys Acta 1863:1799–1805
35. Han D, Tan H, Sun C et al (2018) Dysfunctional $Na_V1.5$ channels due to SCN5A mutations. Exp Biol Med (Maywood) 243:852–863
36. Hyltén-Cavallius L, Iepsen EW, Wewer Albrechtsen NJ et al (2017) Patients with long-QT syndrome caused by impaired hERG-encoded Kv11.1 potassium channel have exaggerated endocrine pancreatic and incretin function associated with reactive hypoglycemia. Circulation 135:1705–1719

37. Sanguinetti MC (2010) HERG1 channelopathies. Pflugers Arch 460:265–276
38. Vandenberg JI, Perry MD, Perrin MJ et al (2012) hERG K(+) channels: structure, function, and clinical significance. Physiol Rev 92:1393–1478
39. He S, Moutaoufik MT, Islam S et al (2020) HERG channel and cancer: a mechanistic review of carcinogenic processes and therapeutic potential. Biochim Biophys Acta Rev Cancer 1873: 188355
40. Drew BJ, Ackerman MJ, Funk M et al (2010) Prevention of torsade de pointes in hospital settings: a scientific statement from the American Heart Association and the American College of Cardiology Foundation. Circulation 121:1047–1060
41. Priest BT, Bell IM, Garcia ML (2008) Role of hERG potassium channel assays in drug development. Channels (Austin) 2:87–93
42. Garrido A, Lepailleur A, Mignani SM et al (2020) hERG toxicity assessment: useful guidelines for drug design. Eur J Med Chem 195:112290
43. Sachse R, Dondapati SK, Fenz SF et al (2014) Membrane protein synthesis in cell-free systems: from bio-mimetic systems to bio-membranes. FEBS Lett 588:2774–2781
44. Thoring L, Wüstenhagen DA, Borowiak M et al (2016) Cell-free systems based on CHO cell lysates: optimization strategies, synthesis of "Difficult-to-Express" proteins and future perspectives. PloS One 11:e0163670
45. Brödel AK, Raymond JA, Duman JG et al (2013) Functional evaluation of candidate ice structuring proteins using cell-free expression systems. J Biotechnol 163:301–310
46. Brödel AK, Sonnabend A, Roberts LO et al (2013) IRES-mediated translation of membrane proteins and glycoproteins in eukaryotic cell-free systems. PloS One 8:e82234
47. Stech M, Brödel AK, Quast RB et al (2013) Cell-free systems: functional modules for synthetic and chemical biology. Adv Biochem Eng Biotechnol 137:67–102
48. Lee SY, Lee A, Chen J et al (2005) Structure of the KvAP voltage-dependent K+ channel and its dependence on the lipid membrane. Proc Natl Acad Sci U S A 102:15441–15446
49. Thompson MJ, Baenziger JE (2020) Ion channels as lipid sensors: from structures to mechanisms. Nat Chem Biol 16:1331–1342
50. Mager MD, Melosh NA (2007) Lipid bilayer deposition and patterning via air bubble collapse. Langmuir 23:9369–9377
51. Jain NK, Barkowski-Clark S, Altman R et al (2017) A high density CHO-S transient transfection system: comparison of ExpiCHO and Expi293. Protein Expr Purif 134:38–46
52. Muraoka T, Noguchi D, Kasai RS et al (2020) A synthetic ion channel with anisotropic ligand response. Nat Commun 11:2924
53. Stamatis C, Farid SS (2021) Process economics evaluation of cell-free synthesis for the commercial manufacture of antibody drug conjugates. Biotechnol J 16:e2000238

Adv Biochem Eng Biotechnol (2023) 186: 121–140
https://doi.org/10.1007/10_2023_219

Published online: 13 June 2023

Progresses in Cell-Free In Vitro Evolution

Kaito Seo, Katsumi Hagino, and Norikazu Ichihashi

Contents

1 Introduction 122
2 Directed Evolution with Cell-Free Systems 123
2.1 Ribosome Display 123
2.2 Nucleic Acid Display 126
2.3 In Vitro Compartmentalization 129
3 Undirected Evolution with Cell-Free Systems 131
3.1 RNA-Based Darwinian Evolution 131
3.2 DNA-Based Darwinian Evolution 134
4 Future Perspective 134
References 136

Abstract Biopolymers, such as proteins and RNA, are integral components of living organisms and have evolved through a process of repeated mutation and selection. The technique of "cell-free in vitro evolution" is a powerful experimental approach for developing biopolymers with desired functions and structural properties. Since Spiegelman's pioneering work over 50 years ago, biopolymers with a wide range of functions have been developed using in vitro evolution in cell-free systems. The use of cell-free systems offers several advantages, including the ability to synthesize a wider range of proteins without the limitations imposed by cytotoxicity, and the capacity for higher throughput and larger library sizes than cell-based

K. Seo and K. Hagino
Department of Life Science, Graduate School of Arts and Science, The University of Tokyo, Tokyo, Japan

N. Ichihashi (✉)
Department of Life Science, Graduate School of Arts and Science, The University of Tokyo, Tokyo, Japan

Komaba Institute for Science, The University of Tokyo, Tokyo, Japan

Universal Biology Institute, The University of Tokyo, Tokyo, Japan
e-mail: ichihashi@bio.c.u-tokyo.ac.jp

evolutionary experiments. In this chapter, we provide a comprehensive overview of the progress made in the field of cell-free in vitro evolution by categorizing evolution into directed and undirected. The biopolymers produced by these methods are valuable assets in medicine and industry, and as a means of exploring the potential of biopolymers.

Graphical Abstract

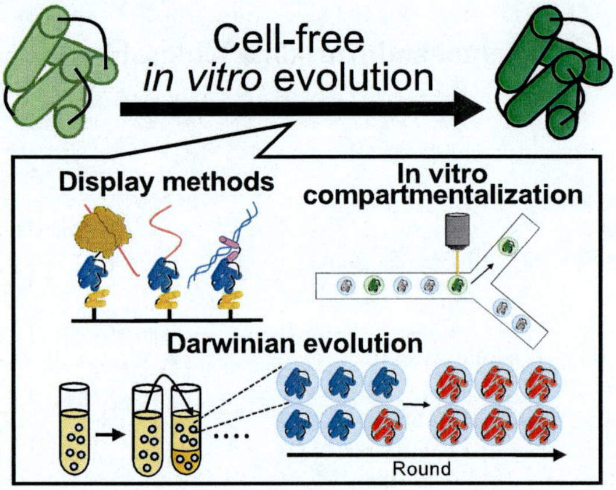

Keywords Darwinian evolution, Directed evolution, In vitro selection, Protein engineering

1 Introduction

The modification of biopolymers, including proteins and nucleotides, to achieve a desired activity and structure is one of the major challenges in biochemical engineering and biotechnology [1–3]. A rational approach to this task is to modify individual monomers (e.g., amino acid residues for proteins) based on sequence and structural information [4–6]. However, despite recent advances such as the development of AlphaFold2 [7], our understanding of the relationship between sequence, structure, and function remains incomplete [8, 9].

The biopolymers found in nature are the result of evolution rather than deliberate design. Through a process of repeated mutation and selection over many generations, these molecules have become highly adapted to their host species and have acquired substantial activity. The evolutionary process has inspired researchers to perform similar experiments with this selection system in the laboratory. In the 1960s, Spiegelman's group performed evolutionary experiments on RNA molecules [10]. This work was later extended to protein evolution by Barry Hall in 1978, who applied specific selective pressures to *Escherichia coli* [11]. Subsequently, Smith and Arnold established cell-based "directed evolution" of proteins using filamentous phages and bacteria, respectively [12, 13]. These advances in evolutionary approaches have led to the development of proteins with improved activity over their naturally occurring counterparts, as well as enzymes with novel activities [14, 15].

The traditional directed evolution approach uses natural cells, such as bacteria, for protein expression. However, this method has certain limitations that can hinder the evolution of cytotoxic proteins, as improved activity can lead to the death of the host cells [16–18]. In addition, the step of introducing mutagenized genes into a host cell limits the size of the library and reduces the overall throughput of the process. To overcome these limitations, cell-free technology has been proposed as a solution [19]. This approach eliminates the problem of cytotoxicity and can be combined with microfluidics to achieve a high throughput experimentation with a library size of up to 10^{13} [20]. These advantages make cell-free technology a promising strategy for engineering biopolymers with desired functions and structures.

In the field of cell-free in vitro evolution, one of the major challenges is to establish a link between genotype and phenotype. In the case of directed evolution of proteins, the phenotype is produced by the proteins, while it is not the proteins that should be selected, but the DNA or mRNA that encodes the information of the proteins (i.e., the genotype). In traditional evolutionary experiments using natural cells, this link is easy to make because the selected cell that expresses the desired phenotype contains the DNA of the desired proteins. In a cell-free system, however, this linkage requires special techniques, such as the formation of complexes with proteins and the original mRNA, or in vitro compartmentalization. The development of a new method to establish the link between genotype and phenotype has been a crucial aspect of the history of cell-free in vitro evolution.

In this chapter, we provide an overview of the methods used in the field of cell-free in vitro evolution. We divide these methods into two categories: directed and undirected evolution. Directed evolution focuses on the artificial selection of proteins and polynucleotides with a desired function. In undirected evolution, polynucleotides and proteins with greater replicability are selected "naturally" according to the Darwinian principle. We also discuss the advantages and disadvantages of each method and possible future directions for cell-free in vitro evolution (Table 1).

2 Directed Evolution with Cell-Free Systems

2.1 *Ribosome Display*

The method of ribosome display establishes a link between phenotype and genotype by forming a protein/ribosome/mRNA complex, selecting proteins with the desired function and mRNAs encoding its information from a random library of sequences. The concept of ribosome display was first introduced by Tuerk and Gold [21]. The

Table 1 Comparison of evolution method with cell-free system

Method	Genotype–phenotype linkage	Library size	Evolvable phenotype	Advantage	Disadvantage	Reference
Ribosome display	Ribosome/mRNA/protein	~10^{13}	Binding interactions, proteins	Established method	Low complex stability	[23]
mRNA display	mRNA/protein	~10^{13}	Binding interactions, proteins	More stable than ribosome display and unaffected by ribosome	Low preparation efficiency of conjugated mRNA	[42, 43]
cDNA display	cDNA/mRNA/protein	~10^{13}	Binding interactions, proteins	More stable than mRNA display and allows post-translational modifications	Complicated experimental procedure	[49]
CIS display	DNA/protein	~10^{13}	Binding interactions, proteins	One-pot preparation and RNase-insensitive	Low stability (~2 days) due to non-covalent bonding	[54]
CAD display	DNA/protein	~10^{7}	Binding interactions, proteins	One-pot preparation and RNase-insensitive, stable due to covalent bond	Small library size	[55]
IVC	Depends on experiments (beads, biotin, etc.)	~10^{10}	Fluorescent proteins, catalytic enzymes combined with fluorescent activities	Ultra-high throughput, wide range of enzyme activities	FACS setup, microfluidic device, high level of experimental technique	[56, 58, 59, 65–67]
Liposome display	Liposome/protein	~10^{7}	Transmembrane protein	Transporters can evolve	FACS setup, low synthesis efficiency of liposomes	[75, 102]
Darwinian evolution	Gene population	~10^{10}	Polymerase related protein	Evolution just by serial-transfer	IVC technique Parasite may spawn	[80, 92]

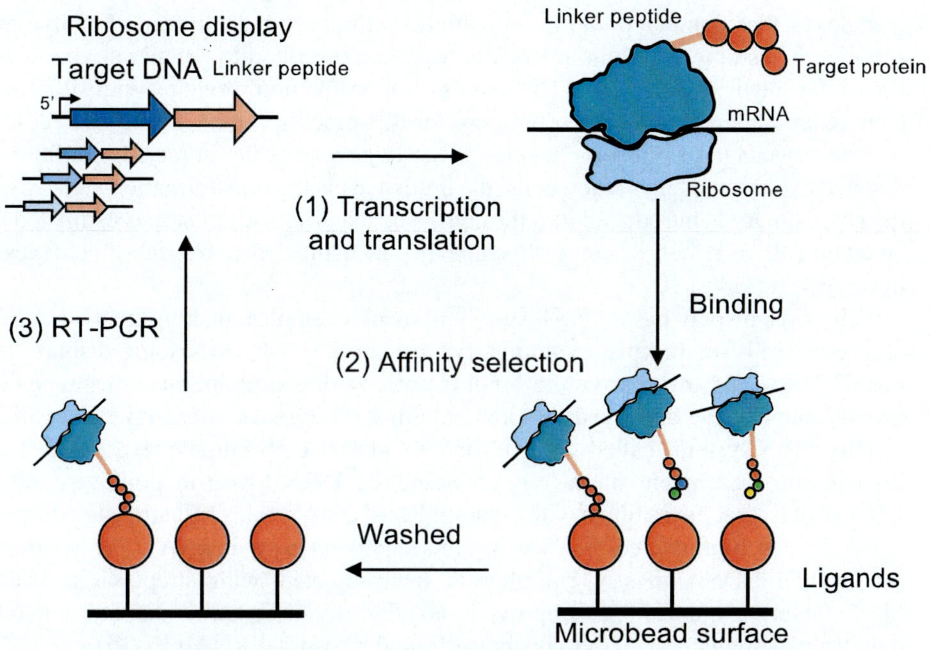

Fig. 1 Ribosome display. Schematic representation of the ribosome display. A round of the selection process consists of four steps. (1) The target DNA library is transcribed and translated in vitro using a cell-free system. (2) Protein/ribosome/mRNA complexes are selected by their affinity. The complex is bound to ligands by the affinity of the expressed proteins. Washing with buffer leaves the complex with strong affinity (affinity selection). (3) Target DNA encoding the high-affinity protein is obtained by reverse transcription and PCR

prototype of the method was reported in 1994, in which only the epitope sequence of a short peptide was selected [22]. In 1997, Hanes and Plückthun introduced the selection of a fully translated protein stalled on the ribosome [23].

Ribosome display is a process that consists of three steps: library construction, cell-free translation of a target protein, and selection based on protein affinity (Fig. 1). The first step involves the generation of a random DNA library using methods such as error-prone PCR and DNA shuffling [20, 24, 25]. This DNA library encodes various mutants of the target gene, which is fused to a downstream linker region and does not contain a stop codon; the coding region continues to the 3′-end of the DNA. In the second step, the libraries are transcribed and translated in a cell-free gene-expression system, resulting in the production of protein variants with varying activities. When the ribosome reaches the 3′-end of the mRNA, it stalls and forms an mRNA/ribosome/target protein complex. This complex is stabilized at a low temperature and by increasing the magnesium concentration. In some cases, the use of the toxin ricin A has been reported to enhance ribosome stalling and allow the reaction to proceed at room temperature [26]. The third step in ribosome display involves the selection of protein/ribosome/mRNA complexes that bind to a target ligand stabilized on beads or a column. This selection process enriches the complexes that contain proteins with higher affinity for the ligand. Finally, the complex is disrupted, and the mRNA is reverse transcribed to regenerate the DNA library for the next generation. The process is repeated until proteins with sufficient activity are obtained. A detailed protocol for this process can be found here [27].

One advantage of ribosome display is the large size of the library. In traditional directed evolution using natural cells, the bottleneck is the transformation step where the DNA library is introduced into the cells. The maximum efficiency is estimated to be around 10^9 [28, 29], whereas ribosome display using cell-free systems achieves a library size of up to 10^{13}.

Ribosome display has a long history and many examples, making it an attractive choice for cell-free in vitro evolutionary methods [30, 31]. Ribosome display has mainly been used to improve the binding activity of a protein, but other types of activity can also be improved in some experiments. One such example is ligation activity. The system, called MDRE (mRNA-dsDNA-ribosome-enzyme) display, involves the attachment of mRNA encoding T4 DNA ligase to double-stranded DNA (dsRNA) at its terminus with a partial RNA/DNA duplex. The translated DNA ligase ligates biotinylated-dsDNA in solution to its own mRNA. The resulting biotinylated-dsDNA/mRNA complex is then selected with streptavidin beads [32]. Other examples of improved activity by ribosome display include dihydrofolate reductase, sialyltransferase II, and ribosomal RNA [33–35].

Several cell-free gene expression systems have been used for ribosome display. In many cases, prokaryotic *E. coli* S30 extract has been used [30]. The cell extract usually contains significant RNase activity, which can degrade the mRNA/ribosome/protein complex. This problem can be mitigated by using an RNase I mutant strain [36]. Another option is the reconstituted *E. coli* translation system, in which all factors have been purified individually, circumventing the problems of some factors in the lysate inhibiting the selection process [37]. In addition, cell extracts from eukaryotic systems such as rabbit reticulocyte, wheat germ, and yeast have been used for the expression of some proteins [38–40].

Ribosome display also has limitations, including the instability of the mRNA/ribosome/protein complex. This instability results in a limited fraction of the complex being subjected to the selection process [41]. The instability also limits the selection pressure that can be applied. The size of the translated protein is also limited, as large proteins may interact with the ribosome and fold into an unexpected state [41].

2.2 *Nucleic Acid Display*

To overcome the problem of instability of the mRNA/ribosome/protein complex in ribosome display, methods have been developed to establish a direct link between the mRNA and the target protein. These methods, known as mRNA display and cDNA display, fall under the broader category of nucleic acid display in this chapter.

In 1997, the groups of Szostak and Yanagawa independently developed a method known as mRNA display (in vitro virus) [42, 43]. This technique consists of the following five steps, as shown in Fig. 2a (left). In the first step, mRNA encoding a mutagenized target protein is synthesized from a DNA library by in vitro transcription. In the second step, puromycin, an antibiotic and a tyrosyl-tRNA analog [44], is conjugated to the 3′-end of the mRNA. In the third step, the target protein is translated from the mRNA-puromycin conjugate by cell-free translation. During this process, the C-terminus of the nascent protein at the P-site of the ribosome covalently binds to puromycin, which has entered the A-site, to form an mRNA/protein conjugate (Fig. 2b) [42]. In the fourth step, mRNA/peptide conjugates are selected by methods similar to ribosome display. Finally, in the fifth step, the mRNA components are converted into DNA by reverse transcription and PCR.

The basic framework of mRNA display is comparable to that of ribosome display and has a similarly large library size ($\sim 10^{13}$) [42]. However, the stability of the covalent mRNA/peptide complex in mRNA display is greater than that of the non-covalent protein/ribosome/mRNA complex in ribosome display [45]. Furthermore, mRNA display can express larger target proteins (~300 AA) than ribosome display because it prevents steric interactions between newly synthesized proteins and ribosomes [41]. As a practical application of this method, Szostak's group has used immunoprecipitation as a selection method to engineer myc epitopes [42]. This technique has also been used to generate ATP or streptavidin aptamers and artificial RNA ligases from random sequences [46–48].

One of the persistent drawbacks of mRNA display is low mRNA stability. While covalent binding improves the stability of mRNA/protein conjugates, mRNA is still susceptible to degradation by RNA-degrading enzymes in cell-free lysates. To overcome this problem, Hushimi et al. developed cDNA display, in which a cDNA/mRNA/peptide complex is formed by performing a reverse transcription reaction using the mRNA/peptide complex as a template [49] (Fig. 2a, right). In contrast to mRNA display, this method involves covalent attachment of cDNA to a target protein. Due to the high stability of cDNA, this complex can be protected from degradation by RNases [50]. The stability of cDNA/protein complex allows the selection of peptides in a serum containing high levels of RNase [50]. As an alternative ligation method, Fujimoto et al. developed a photocrosslinking technique between mRNA and puromycin using 3-cyanovinylcarbazole (cnvK). This process was completed in 1 min with UV irradiation, avoiding the previous lengthy ligation process (15–40 h) [51–53]. Another advantage of these new ligation methods is that there is no need for mRNA purification after the ligation step, resulting in a higher yield of mRNA-puromycin linker.

The mRNA and cDNA displays have certain limitations compared to the ribosome display, one of which is the need for multi-step experimental procedures prior to the selection process. To overcome this, CIS display and CAD display have been developed as one-pot reactions to obtain nucleic acid–protein complexes (Fig. 2c). McGregor et al. developed CIS display, a DNA/protein display method in which the repA gene, CIS sequence, and ori sequence are attached downstream of the target gene [54]. During the one-pot transcription/translation reaction, RNA polymerase

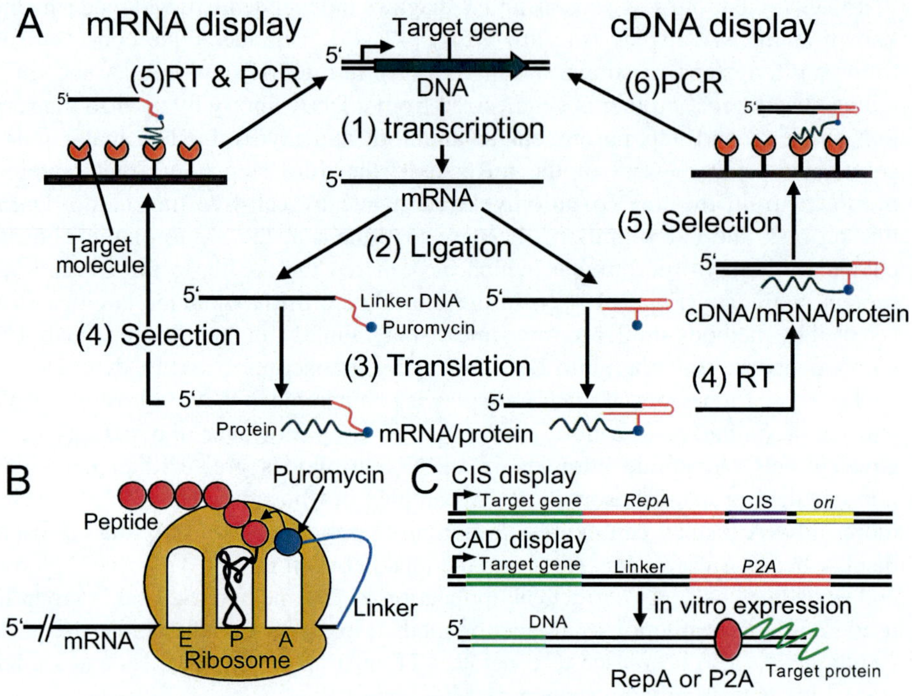

Fig. 2 Nucleic acids display. Schematic representation of four approaches to nucleic acids display. (**a**) The experimental flow of mRNA display and cDNA display. mRNA display (left) consists of five steps. (1) In vitro transcription of the target genes. (2) Puromycin linker ligation. (3) Translation reaction in a cell-free translation system. (4) Affinity selection with target molecules. (5) Recovery of RNA, reverse transcription reaction, and PCR. cDNA display (right) performs formation of cDNA/mRNA/protein complex by reverse transcription before selection. (**b**) Mechanism of covalent binding of puromycin to the nascent peptide in the ribosome. Puromycin bound to the linker DNA enters the A site when approached by translating ribosomes and covalently binds to the C-terminus of the nascent peptide at the P site. (**c**) Schematic representation of the CIS and CAD displays. Target protein fused to RepA or P2A binds to DNA encoding itself after in vitro expression. RepA binds DNA non-covalently, whereas P2A binds DNA covalently

transcribes the target gene and the repA gene and then stalls at the CIS sequence, allowing the target/RepA fusion protein translated from the mRNA to attach to the stalled RNA polymerase with DNA. The RepA region of the fused protein then binds to the ori sequence on the DNA to form a DNA/mRNA/protein complex.

The CAD display developed by Marvik et al. is a simpler method, in which DNA encodes the bacteriophage P2-derived endonuclease P2A genes [55]. The P2A gene has the ability to covalently bind to the 5′ phosphate of its own DNA, and thus in the one-pot transcription and translation reaction, the P2A gene forms a DNA/protein conjugate immediately after transcription. Because the CAD display is covalent, it is more stable than the non-covalent CIS display. Currently, the CAD display has a library size of 10^7 and an enrichment factor of approximately 10^2, which is lower than the CIS display with a library size of 10^{13} and an enrichment factor of 10^3 [55].

In conclusion, each of the four display methods has its own set of advantages and disadvantages, and it is recommended to choose the most suitable method based on the type of protein to be engineered and the required library size.

2.3 In Vitro Compartmentalization

The concept of using the in vitro compartment (IVC) as a means of establishing a genotype–phenotype link was first proposed and demonstrated by Tawfik and Griffiths [56]. In this method, the function of the target protein (i.e., the phenotype) is assessed within a single compartment. To obtain a clear genotype–phenotype association, it is essential that each compartment contains a small number of copies of the target gene (ideally only one). If a compartment contains multiple copies of the gene with different mutations, the effect of these mutations on the phenotype will be diluted, reducing the efficiency of selection.

There are several strategies for selecting the desired target function in the IVC method (Fig. 3). Tawfik and Griffiths first used water-in-oil droplets as compartments, encapsulating a DNA methyltransferase Hae III gene library and a cell-free gene expression system. During the reaction, the expressed methyltransferase methylates its own DNA. The droplets were then collected and the non-methylated DNA was degraded, enriching for the more methylated (i.e., more active) mutant genes [56]. Later, Doi and Yanagawa developed a system known as STABLE, in which DNA and translated proteins were linked by microbeads and antibodies in individual compartments [57]. Sepp et al. used a cell sorter for selection, in which, each bead linked to DNA and proteins was subjected to a cell sorter after the droplets were broken to collect those with higher fluorescence, which represents higher target protein function [58, 59].

This fluorescence-based selection method is applicable to a wide range of targets because fluorescence can be used to monitor various functions. As another method to use a cell sorter, Bernath et al. reported re-emulsification of w/o droplets to form w/o/w droplets to be applicable to a cell sorter [60]. Other selection methods include mineral-based selection, where enzymes that produce minerals are selected based on the scattered light produced by the minerals [61].

In addition to the above, there are two other methods for linking phenotype and genotype within droplets: covalent DNA display and SNAP display. Covalent DNA display was developed by Bertschinger et al. [62]. In this system, the DNA contains a target gene fused to DNA methyltransferase and a crosslinking sequence containing 5-fluoro-2′deoxycytidine. The translated methyltransferase fusion protein itself catalyzes the crosslinking with the crosslinking sequence. The resulting DNA/protein complex is then selected for its binding activity to a ligand. In the SNAP display developed by Hollfelder's group, the DNA encoding the fusion protein of interest and an O^6-alkylguanine DNA alkyltransferase (AGT) are translated in a cell-free system within droplets [63, 64]. The DNA is labeled with benzylguanine (BG). The translated AGT catalyzes the formation of covalent

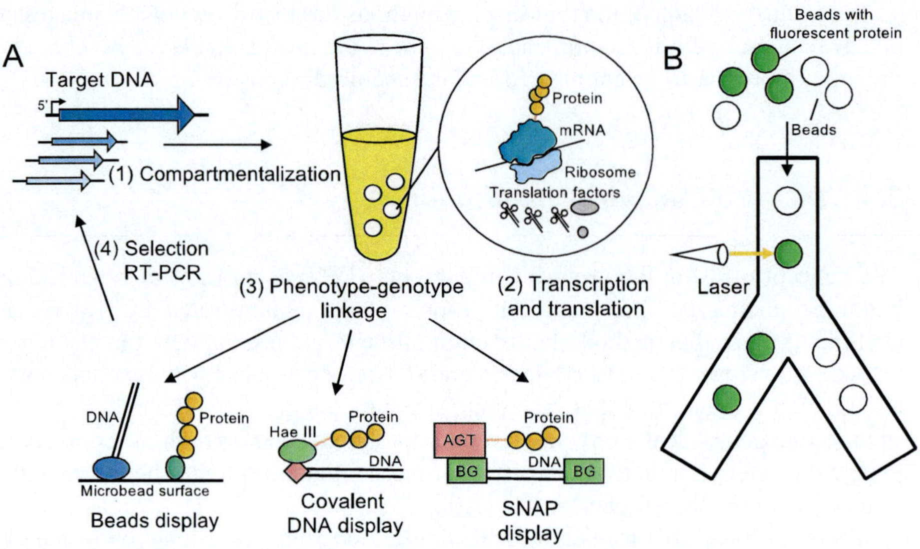

Fig. 3 In vitro compartmentalization. Schematic representation of four approaches using in vitro compartmentalization. (**a**) The experimental flow of the evolution method using w/o droplets. (1) The library of target DNA is compartmentalized inside the droplets with a cell-free system. (2) The target DNA is transcribed and translated inside the droplet. (3) DNA and protein are linked inside the w/o droplets by various methods. Schemes of three linking methods; beads display, covalent DNA display, and SNAP display are shown. In the beads display, the DNA and protein are attached to the surface of the same beads. In the covalent DNA display, the target protein is fused to a methyltransferase that recognizes and covalently binds to the cross-linking sequence of the DNA containing 5-fluoro-2′deoxycitidine. In the SNAP display, the target protein is fused to O^6-alkylguanine DNA alkyltransferase (AGT), which binds to its substrate benzylguanine (BG) labeled on the original DNA. (4) Target proteins with higher activity are selected, and the fused DNA is recovered by reverse transcription and PCR. (**b**) Schematic representation of compartment sorting. Each bead (or liposome, w/o/w droplet) passes through a laser and is sorted according to the fluorescence, which reflects the activity of a target protein

bonds with BG in the DNA termini. The DNA/target protein-AGT complex is then selected by affinity selection of target proteins.

The water-in-oil droplets used in the above studies were prepared by vigorously mixing a reaction solution in oil and surfactants. This approach resulted in droplets with a large size distribution, which could reduce reaction and selection efficiency. More homogeneous droplets have been generated using microfluidic techniques [65–67]. The use of microfluidics also reduces the amount of cell-free system used for an experiment, thereby reducing cost [68]. The combination of microfluidics and droplet sorting has been shown to increase the speed and efficiency of selection [69]. In addition, microfluidics can be coupled with other evaluation methods, such as mass spectrometry [70].

Giant unilamellar vesicles, or liposomes, have also been reported as an IVC method. Using a cell sorter, Uyeda et al. were able to develop an aminoacyl-tRNA synthetase that showed enhanced binding of N-benzyloxycarbonyl-l-lysine

[71]. Meanwhile, Murase et al. found a mutant 16S rRNA that retained functional activity even in the absence of post-translational modifications in a high salt buffer [72]. Recently, liposome-based IVC was used to select for the self-replication activity of DNA encoding DNA replication factors of bacteriophage phi29 [73].

The use of liposome-based IVC is crucial for the directed evolution of membrane proteins. When expressed in a liposome-containing cell-free gene expression system, a certain fraction of membrane proteins are inserted into the membrane [74]. In a study by Fujii et al., a random DNA library of α-hemolysin, a channel membrane protein of *Staphylococcus aureus*, was expressed in liposomes. The liposomes also contain a peptide tag that binds to a fluorescent molecule located outside the liposomes. The efficiency of hemolysin in forming pores on the liposome membrane was monitored by the influx of the fluorescent molecule into the liposomes. A cell sorter was then used to collect highly fluorescent liposomes, which were expected to contain hemolysin mutant genes with higher pore-forming activity, to obtain the mutant gene sequences [75].

The versatility of using IVC for directed evolution is a significant advantage over other display methods, such as ribosome display or mRNA display. With IVC, there are fewer restrictions on the activity of the target protein, making it a more flexible method. If the activity of the target protein is linked to the selection process, e.g. by fluorescence or biotinylation, the protein activity can be monitored and selected by cell sorting or affinity binding [76]. There remains a significant opportunity for further research to expand the range of proteins that can be improved using IVC.

A disadvantage of using IVC for directed evolution is the smaller library size (10^4~10^{10}) compared to the other display methods (10^{13}) [77]. The library size mainly depends on the number of compartments that can be prepared. The simple mixing method can produce a large number of water-in-oil droplets, typically 10^5~10^8, while the use of microfluidics yields a smaller number [78]. The preparation of liposomes, which is more complex, typically results in a much smaller number of compartments. The development of faster and more efficient methods for preparing compartments could increase the possible library size for IVC-based directed evolution. Another drawback of using IVC for directed evolution is the need for specialized and often expensive equipment such as a cell sorter and a microfluidic device. The development of more cost-effective selection methods could increase the utility of this directed evolution approach.

3 Undirected Evolution with Cell-Free Systems

3.1 RNA-Based Darwinian Evolution

In the previous Sect. 2, we presented many examples of directed evolution, where DNA or RNA was selected based on its respective protein activity, and thus the evolution was "directed" to improve the target protein activity. However, the evolutionary scheme is different in natural organisms, which evolved based on their own self-replicating ability (fitness) according to the Darwinian principle. In this evolutionary scheme, organisms that can produce more copies of themselves (i.e., offspring) dominate the population. This natural type of evolution is referred to in this chapter as undirected or Darwinian evolution.

In 1967, the first cell-free in vitro Darwinian evolution was performed by Spiegelman's group. In this experiment, the RNA genome of bacteriophage Qβ was incubated with the replication enzyme (Qβ replicase) of the same bacteriophage to replicate the RNA. The replication reaction was repeated for many generations through the serial dilution cycle [10]. During the replication for many generations, spontaneous mutations (especially deletions) were introduced into the RNA, and short mutant RNAs that can replicate faster evolved. In this experiment, however, the RNA replication enzyme is not a target of evolution because the enzyme was not expressed from the RNA but was supplied as a purified protein.

In 2013, we extended Spiegelman's system by introducing the reconstituted cell-free translation system [79] to achieve Darwinian evolution based on translation-coupled RNA replication in the IVC [80] (Fig. 4a). In contrast to Spiegelman's experiment, in which shorter RNAs evolved, the RNAs retained their original size but increased their replication efficiency. In this evolutionary experiment, another biologically relevant phenomenon is observed: the appearance of parasitic RNAs that have lost the RNA replicase genes but can replicate by exploiting the replicase translated from another intact RNA (host RNA). Once such parasitic RNAs appeared, the population dynamics of parasite and host RNA oscillated [81] (Fig. 4b), which is the same phenomenon sometimes observed between predator and prey in nature [82–85]. This evolutionary experiment continued, and at the latest stage of 240 rounds (1,200 h), we found that the RNA had diversified into at least five distinct lineages, including three host and two parasite lineages, four of which form an interdependent replicator network [86] (Fig. 4c). The Darwinian evolutionary system of RNA may be a useful experimental model for understanding the evolutionary scenario of early life forms. In addition, we have obtained various mutant RNA replicases, including those that have acquired different template specificity [86] and also relaxed substrate specificity [87] through this long-term evolution.

The advantage of Darwinian undirected evolution over directed evolution is its simple methodology. The process required for Darwinian evolution is simply to repeat the serial dilution cycle. Mutations are introduced spontaneously during replication by replication error, and selection for more replicable RNA occurs automatically according to the Darwinian principle. Because of this simplicity, it is relatively easy to set up an automated experiment [88]. The disadvantage is the uncontrollability of the improved character. Since Darwinian evolution is "undirected," we cannot determine which activity should be improved during evolution. It should depend on the rate-limiting step for replication. For example, in our previous RNA evolution, the activity of RNA as a template (i.e., the activity to be recognized as a replication template by the replicase) was improved in the initial stage of evolution, but the template specificity of the replicase evolved in the later stage. This uncontrollability is not suitable if one wants to improve a specific activity of a

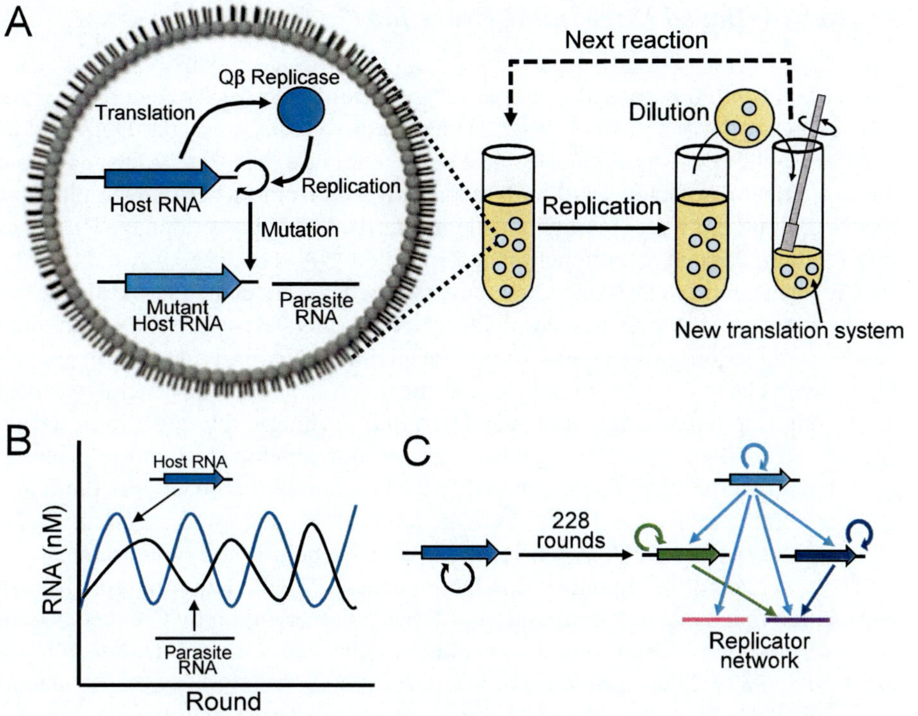

Fig. 4 RNA-based Darwinian evolution. Schematic representation of RNA-based Darwinian evolution. (**a**) Darwinian evolution experiments of a translation-coupled RNA replication system through a serial dilution cycle in an IVC. In the w/o emulsion, an RNA encoding the Qβ replicase (host RNA) self-replicates by expressing the replicase in a cell-free translation system. Mutant host RNAs and another small RNAs that do not encode the replicase gene (parasite RNA) were produced by introducing deletion through RNA self-replication. This reaction was repeated for many generations through a serial transfer method. (**b**) RNA population dynamics in Darwinian evolution experiments. The population dynamics of the host and parasite RNAs exhibited oscillatory dynamics often observed in nature in predator–prey or host–parasite relationships. (**c**) Diversification of the host and parasite RNAs. As a result of 228 rounds of Darwinian evolution experiments, one type of host RNA was diversified into three types of the host RNAs and two types of the parasite RNAs. Four of these RNAs formed an interdependent replicator network

protein. Another disadvantage of Darwinian evolution is the limited number of target proteins. In the Darwinian evolutionary scheme, only genes that contribute to self-replication evolve. For the RNA-based system, we have currently succeeded in constructing an RNA replication system that encodes only two genes due to the difficulty in designing the RNA structure [89]. To expand the encoded genes, a DNA-based self-replication system may be a better choice.

3.2 DNA-Based Darwinian Evolution

Forster and Church proposed a simple self-replicating DNA system combined with gene expression in a cell-free system. The system contains a circular DNA and two enzymes, phi29 DNA polymerase and Cre recombinase [90]. Phi29 DNA polymerase is a strand-displacing DNA polymerase derived from bacteriophage phi29 and Cre recombinase is a recombination enzyme derived from bacteriophage P1 that can catalyze homologous recombination at a specific DNA sequence known as the loxP site. When a circular DNA encoding these two genes is incubated in a cell-free gene expression system, the expressed phi29 DNA polymerase initiates DNA polymerization from random primers attached to the circular DNA. Polymerization proceeds for many cycles due to the strand-displacement activity of the polymerase, generating a long, repetitive single-stranded DNA (i.e., rolling-circle replication occurs) (Fig. 5). The polymerase then synthesizes the complementary strand to produce a long, repetitive dsDNA. The expressed Cre recombinase then catalyzes the homologous recombination at the loxP sites in the long dsDNA to reproduce circular DNA, allowing the circular DNA to self-replicate with the help of the two proteins.

The first attempt to implement the DNA replication scheme proposed by Forster and Church failed due to the inhibition of DNA polymerization by Cre recombinase. However, through several rounds of evolution with artificial amplification, a mutant DNA and phi29 DNA polymerase resistant to Cre recombinase were obtained [91]. This led to the realization of a transcription/translation-coupled DNA replication system (TTcDR system) in a cell-free system [91]. Subsequently, Darwinian evolution experiments were performed through a serial dilution cycle in an IVC for 30 rounds (480 h), resulting in the evolution of DNA mutants with higher replication activity [92] (Fig. 5, right). This is another experimental model of Darwinian evolution.

The advantage of the DNA replication scheme is that it is easy to increase the number of genes encoded by DNA. In fact, Libicher et al. successfully achieved the transcription, translation, and replication of DNA encoding translation factors of the *E. coli* translation system [93]. The DNA replication system has a high potential to encode all RNA and protein factors involved in the cell-free gene expression system into self-replicating DNA to make the system self-sustaining. However, the feasibility of replicating such long DNA using the rolling-circle replication remains the next major challenge.

4 Future Perspective

In cell-free directed evolution, several techniques have been developed and many proteins have been improved. However, there is a possible future direction to further improve the selection efficiency. Currently, cell-free technology is limited by its translation efficiency [94]. Improving the gene expression activity of cell-free

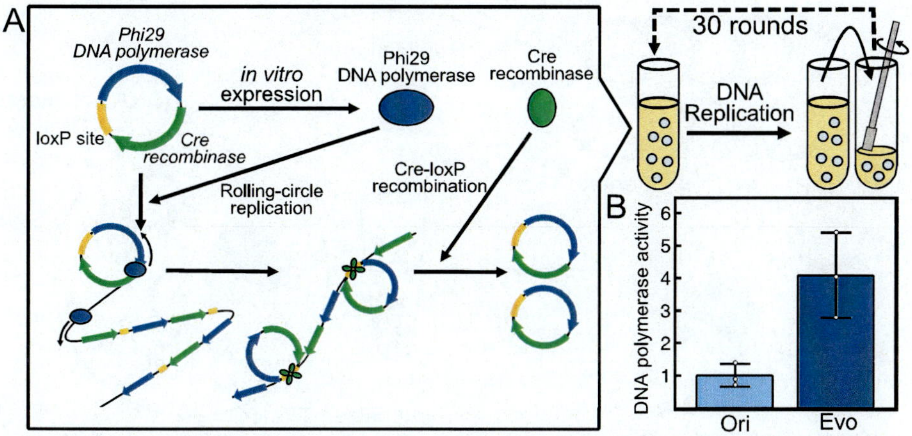

Fig. 5 DNA-based Darwinian evolution. (**a**) Schematic representation of the transcription/translation-coupled DNA replication system. Circular DNA self-replicates through the expression of two self-encoded proteins (phi29 DNA polymerase and Cre recombinase). First, phi29 DNA polymerase catalyzes rolling-circle replication of the circular DNA, producing a long strand of DNA containing repetitive sequences. Next, Cre recombinase recognizes the loxP site on the repetitive sequence and circularizes by Cre-loxP recombination. (**b**) The result of the Darwinian evolution. The mutant DNA that obtained after 30 rounds (480 h) of the evolution (Evo) expressed a higher DNA polymerase activity than the original DNA (Ori)

systems could directly increase the efficiency of evolutionary methods [19]. Another important direction is to increase the range of choices available for cell-free gene expression systems. For example, proteins that cannot be expressed in their active form in the commonly used *E. coli* cell-free systems may be expressed in systems derived from other organisms [95]. Directed evolution has already produced several important proteins, including those with pharmaceutical applications [96], and it is expected to continue to improve useful proteins.

The techniques of the undirected Darwinian evolution in cell-free systems are still in their infancy. Currently, the gene expression-coupled Darwinian evolution has been achieved in only two systems: the RNA replication system using Qβ replicase and the DNA replication system using phi29 DNA polymerase and Cre recombinase. An important future direction is to expand the variety of replication schemes. Several DNA replication systems have been implemented in cell-free systems, such as the genomic DNA replication systems of *E. coli* and bacteriophage phi29 [73, 93, 97, 98], and some have been encapsulated in liposomes [73]. Further research is needed, but these natural genomic DNA replication systems would be useful for Darwinian evolution of longer DNA sequences.

Advancing DNA-based Darwinian evolution also requires increasing the number of genes encoded in DNA. Currently, the RNA and proteins required for gene expression in the cell-free system are supplied from an external source, limiting the potential for evolution. However, if all the factors involved in gene expression could be expressed from the DNA itself, the DNA replication system would become self-sustaining when supplied with low-molecular-weight compounds (Fig. 6). In

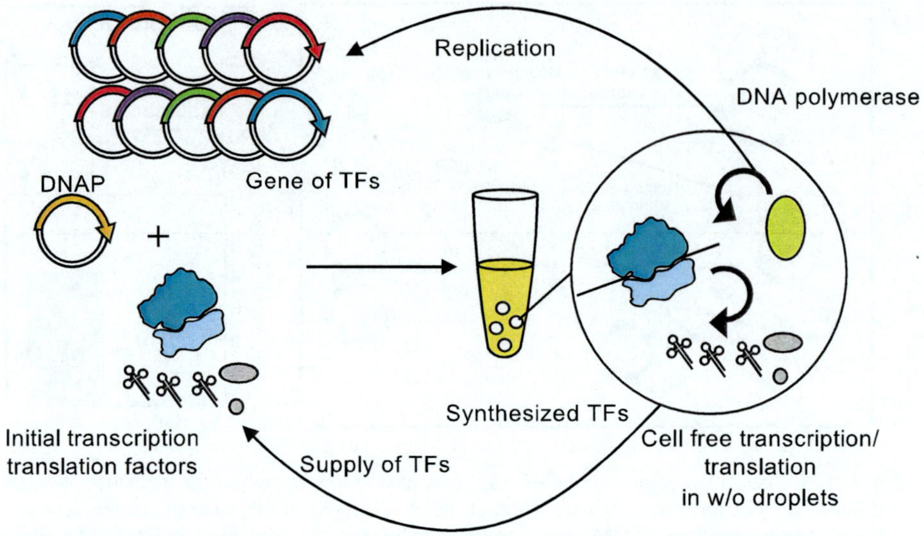

Fig. 6 Perspective of Evolution with a cell-free system. Schematic representation of translation factor (TF) regeneration in a cell-free DNA replication system. DNA encoding TFs and DNA polymerase (DNAP) are encapsulated in w/o droplets with a cell-free system. TFs and DNAP are expressed within the w/o droplets. DNAP replicates the DNA, and the expressed TFs support further gene expression, resulting in continuous transcription/translation/DNA replication. Repeating this process would select mutant TFs and DNAP with higher activities through Darwinian evolution

such a system, all encoded genes such as tRNA, ribosome, aminoacyl-tRNA synthetase (aaRS), and other translation factors could evolve to become more active in vitro. This in vitro evolution of translation factors has the potential to create a specialized in vitro gene expression system that surpasses the capabilities of those found in the cell. Efforts have been made, including by our group, to regenerate translation factors, such as aaRS [99], RNA polymerase [100], and tRNA [101], to construct a self-sustaining cell-free system. The ultimate goal of this effort is to create a completely self-regenerating artificial cell. Such an artificial cell would provide greater control over natural, complex cells and help us understand the boundary between living and non-living systems. The development of a self-regenerating system would provide new technologies and expand our understanding of the fundamental nature of life.

References

1. Arnold FH, Moore JC (1997) Optimizing industrial enzymes by directed evolution. Adv Biochem Eng Biotechnol 58:1–14
2. Cherry JR, Fidantsef AL (2003) Directed evolution of industrial enzymes: an update. Curr Opin Biotechnol 14:438–443
3. Iqbal Z, Sadaf S (2022) Forty years of directed evolution and its continuously evolving technology toolbox: a review of the patent landscape. Biotechnol Bioeng 119:693–724

4. Hellinga HW (1997) Rational protein design: combining theory and experiment. Proc Natl Acad Sci 94:10015–10017
5. Lehmann M, Wyss M (2001) Engineering proteins for thermostability: the use of sequence alignments versus rational design and directed evolution. Curr Opin Biotechnol 12:371–375
6. Porebski BT, Buckle AM (2016) Consensus protein design. Protein Eng Des Sel 29:245–251
7. Jumper J et al (2021) Highly accurate protein structure prediction with AlphaFold. Nature 596: 583–589
8. Hummer AM, Abanades B, Deane CM (2022) Advances in computational structure-based antibody design. Curr Opin Struct Biol 74:102379
9. Arnold FH (2018) Directed evolution: bringing new chemistry to life. Angew Chem Int Ed 57: 4143–4148
10. Mills DR, Peterson RL, Spiegelman S (1967) An extracellular Darwinian experiment with a self-duplicating nucleic acid molecule. Proc Natl Acad Sci 58:217–224
11. Hall BG (1978) Experimental evolution of a new enzymatic function. II. Evolution of multiple functions for EBG enzyme in E. coli. Genetics 89:453–465
12. Smith GP (1985) Filamentous fusion phage: novel expression vectors that display cloned antigens on the virion surface. Science 228:1315–1317
13. Chen K, Arnold FH (1993) Tuning the activity of an enzyme for unusual environments: sequential random mutagenesis of subtilisin E for catalysis in dimethylformamide. Proc Natl Acad Sci 90:5618–5622
14. Bornscheuer UT et al (2012) Engineering the third wave of biocatalysis. Nature 485:185–194
15. Brustad EM, Arnold FH (2011) Optimizing non-natural protein function with directed evolution. Curr Opin Chem Biol 15:201–210
16. De Marco A, Vigh L, Diamant S, Goloubinoff P (2005) Native folding of aggregation-prone recombinant proteins in Escherichia coli by osmolytes, plasmid- or benzyl alcohol-overexpressed molecular chaperones. Cell Stress Chaperones 10:329–339
17. Schlegel S et al (2013) Optimizing heterologous protein production in the periplasm of E. coli by regulating gene expression levels. Microb Cell Factories 12:24
18. Hooe SL, Ellis GA, Medintz IL (2022) Alternative design strategies to help build the enzymatic retrosynthesis toolbox. RSC Chem Biol 3:1301–1313
19. Dodevski I, Markou GC, Sarkar CA (2015) Conceptual and methodological advances in cell-free directed evolution. Curr Opin Struct Biol 33:1–7
20. Packer MS, Liu DR (2015) Methods for the directed evolution of proteins. Nat Rev Genet 16: 379–394
21. Tuerk C, Gold L (1990) Systematic evolution of ligands by exponential enrichment: RNA ligands to bacteriophage T4 DNA polymerase. Science 249:505–510
22. Mattheakis LC, Bhatt RR, Dower WJ (1994) An in vitro polysome display system for identifying ligands from very large peptide libraries. Proc Natl Acad Sci U S A 91:9022–9026
23. Hanes J, Plückthun A (1997) In vitro selection and evolution of functional proteins by using ribosome display. Proc Natl Acad Sci U S A 94:4937–4942
24. Leung D, Chen E, Goeddel D (1989) A method for random mutagenesis of a defined DNA segment using a modified polymerase chain reaction
25. Stemmer WPC (1994) Rapid evolution of a protein in vitro by DNA shuffling. Nature 370: 389–391
26. Zhou J-M, Fujita S, Warashina M, Baba T, Taira K (2002) A novel strategy by the action of ricin that connects phenotype and genotype without loss of the diversity of libraries. J Am Chem Soc 124:538–543
27. Dreier B, Plückthun A (2011) Ribosome display: a technology for selecting and evolving proteins from large libraries. Methods Mol Biol 687:283–306
28. Dower WJ, Miller JF, Ragsdale CW (1988) High efficiency transformation of E. coli by high voltage electroporation. Nucleic Acids Res 16:6127
29. Liu J et al (2018) An improved method of preparing high efficiency transformation Escherichia coli with both plasmids and larger DNA fragments. Indian J Microbiol 58:448

30. Plückthun A (2012) Ribosome display: a perspective. In: Douthwaite JA, Jackson RH (eds) Ribosome display and related technologies: methods and protocols. Springer, New York, pp 3–28
31. Li R, Kang G, Hu M, Huang H (2019) Ribosome display: a potent display technology used for selecting and evolving specific binders with desired properties. Mol Biotechnol 61:60–71
32. Takahashi F et al (2005) Activity-based in vitro selection of T4 DNA ligase. Biochem Biophys Res Commun 336:987–993
33. Bieberich E, Kapitonov D, Tencomnao T, Yu RK (2000) Protein-ribosome-mRNA display: affinity isolation of enzyme-ribosome-mRNA complexes and cDNA cloning in a single-tube reaction. Anal Biochem 287:294–298
34. Takahashi F et al (2002) Ribosome display for selection of active dihydrofolate reductase mutants using immobilized methotrexate on agarose beads. FEBS Lett 514:106–110
35. Hammerling MJ et al (2020) In vitro ribosome synthesis and evolution through ribosome display. Nat Commun 11:1–10
36. Schaffitzel C, Hanes J, Jermutus L, Plückthun A (1999) Ribosome display: an in vitro method for selection and evolution of antibodies from libraries. J Immunol Methods 231:119–135
37. Villemagne D, Jackson R, Douthwaite JA (2006) Highly efficient ribosome display selection by use of purified components for in vitro translation. J Immunol Methods 313:140–148
38. He M (1997) Antibody-ribosome-mRNA (ARM) complexes as efficient selection particles for in vitro display and evolution of antibody combining sites. Nucleic Acids Res 25:5132–5134
39. Alexandrov AN, Alakhov VY, Miroshnikov AI (2002) Wheat germ cell-free translation system as a tool for in vitro selection of functional proteins. Comb Chem High Throughput Screen 5:473–480
40. Gan R, Jewett MC (2016) Evolution of translation initiation sequences using in vitro yeast ribosome display. Biotechnol Bioeng 113:1777–1786
41. Wang H, Liu R (2011) Advantages of mRNA display selections over other selection techniques for investigation of protein–protein interactions. Expert Rev Proteomics 8:335–346
42. Roberts RW, Szostak JW (1997) RNA-peptide fusions for the in vitro selection of peptides and proteins. Proc Natl Acad Sci 94:12297–12302
43. Nemoto N, Miyamoto-Sato E, Husimi Y, Yanagawa H (1997) In vitro virus: bonding of mRNA bearing puromycin at the 3′-terminal end to the C-terminal end of its encoded protein on the ribosome in vitro. FEBS Lett 414:405–408
44. Pestka S (1971) Inhibitors of ribosome functions. Annu Rev Microbiol 25:487–562
45. Newton MS, Cabezas-Perusse Y, Tong CL, Seelig B (2020) In vitro selection of peptides and proteins – advantages of mRNA display. ACS Synth Biol 9:181–190
46. Keefe AD, Szostak JW (2001) Functional proteins from a random-sequence library. Nature 410:715–718
47. Wilson DS, Keefe AD, Szostak JW (2001) The use of mRNA display to select high-affinity protein-binding peptides. Proc Natl Acad Sci 98:3750–3755
48. Seelig B, Szostak JW (2007) Selection and evolution of enzymes from a partially randomized non-catalytic scaffold. Nature 448:828–831
49. Tabuchi I, Soramoto S, Nemoto N, Husimi Y (2001) An in vitro DNA virus for in vitro protein evolution. FEBS Lett 508:309–312
50. Ueno S et al (2012) In vitro selection of a peptide antagonist of growth hormone secretagogue receptor using cDNA display. Proc Natl Acad Sci 109:11121–11126
51. Yoshimura Y, Fujimoto K (2008) Ultrafast reversible photo-cross-linking reaction: toward in situ DNA manipulation. Org Lett 10:3227–3230
52. Mochizuki Y, Suzuki T, Fujimoto K, Nemoto N (2015) A versatile puromycin-linker using cnvK for high-throughput in vitro selection by cDNA display. J Biotechnol 212:174–180
53. Tabata N et al (2009) Rapid antibody selection by mRNA display on a microfluidic chip. Nucleic Acids Res 37:e64–e64
54. Odegrip R et al (2004) CIS display: in vitro selection of peptides from libraries of protein–DNA complexes. Proc Natl Acad Sci 101:2806–2810

55. Reiersen H et al (2005) Covalent antibody display – an in vitro antibody-DNA library selection system. Nucleic Acids Res 33:e10–e10
56. Tawfik DS, Griffiths AD (1998) Man-made cell-like compartments for molecular evolution. Nat Biotechnol 16:652–656
57. Doi N, Yanagawa H (1999) STABLE: protein-DNA fusion system for screening of combinatorial protein libraries in vitro. FEBS Lett 457:227–230
58. Sepp A, Tawfik DS, Griffiths AD (2002) Microbead display by in vitro compartmentalisation: selection for binding using flow cytometry. FEBS Lett 532:455–458
59. Griffiths AD, Tawfik DS (2003) Directed evolution of an extremely fast phosphotriesterase by in vitro compartmentalization. EMBO J 22:24–35
60. Bernath K et al (2004) In vitro compartmentalization by double emulsions: sorting and gene enrichment by fluorescence activated cell sorting. Anal Biochem 325:151–157
61. Bawazer LA et al (2012) Evolutionary selection of enzymatically synthesized semiconductors from biomimetic mineralization vesicles. Proc Natl Acad Sci U S A 109:E1705–E1714
62. Bertschinger J, Grabulovski D, Neri D (2007) Selection of single domain binding proteins by covalent DNA display. Protein Eng Des Sel 20:57–68
63. Stein V, Sielaff I, Johnsson K, Hollfelder F (2007) A covalent chemical genotype-phenotype linkage for in vitro protein evolution. Chembiochem 8:2191–2194
64. Kaltenbach M, Hollfelder F (2012) SNAP display: in vitro protein evolution in microdroplets. Methods Mol Biol 805:101–111
65. Courtois F et al (2008) An integrated device for monitoring time-dependent in vitro expression from single genes in picolitre droplets. Chembiochem 9:439–446
66. Kintses B et al (2012) Picoliter cell lysate assays in microfluidic droplet compartments for directed enzyme evolution. Chem Biol 19:1001–1009
67. Fallah-Araghi A, Baret J-C, Ryckelynck M, Griffiths AD (2012) A completely in vitro ultrahigh-throughput droplet-based microfluidic screening system for protein engineering and directed evolution. Lab Chip 12:882–891
68. Agresti JJ et al (2010) Ultrahigh-throughput screening in drop-based microfluidics for directed evolution. Proc Natl Acad Sci U S A 107:4004–4009
69. Zinchenko A et al (2014) One in a million: flow cytometric sorting of single cell-lysate assays in monodisperse picolitre double emulsion droplets for directed evolution. Anal Chem 86: 2526–2533
70. Diefenbach XW et al (2018) Enabling biocatalysis by high-throughput protein engineering using droplet microfluidics coupled to mass spectrometry. ACS Omega 3:1498–1508
71. Uyeda A et al (2015) Liposome-based in vitro evolution of aminoacyl-tRNA synthetase for enhanced pyrrolysine derivative incorporation. Chembiochem 16:1797–1802
72. Murase Y, Nakanishi H, Tsuji G, Sunami T, Ichihashi N (2018) In vitro evolution of unmodified 16S rRNA for simple ribosome reconstitution. ACS Synth Biol 7:576–583
73. van Nies P et al (2018) Self-replication of DNA by its encoded proteins in liposome-based synthetic cells. Nat Commun 9:1583
74. Kalmbach R et al (2007) Functional cell-free synthesis of a seven helix membrane protein: in situ insertion of bacteriorhodopsin into liposomes. J Mol Biol 371:639–648
75. Fujii S, Matsuura T, Sunami T, Kazuta Y, Yomo T (2013) In vitro evolution of α-hemolysin using a liposome display. Proc Natl Acad Sci 110:16796–16801
76. Markel U et al (2020) Advances in ultrahigh-throughput screening for directed enzyme evolution. Chem Soc Rev 49:233–262
77. Galán A et al (2016) Library-based display technologies: where do we stand? Mol BioSyst 12: 2342–2358
78. Mair P, Gielen F, Hollfelder F (2017) Exploring sequence space in search of functional enzymes using microfluidic droplets. Curr Opin Chem Biol 37:137–144
79. Shimizu Y et al (2001) Cell-free translation reconstituted with purified components. Nat Biotechnol 19:751–755

80. Ichihashi N et al (2013) Darwinian evolution in a translation-coupled RNA replication system within a cell-like compartment. Nat Commun 4:2494
81. Bansho Y, Furubayashi T, Ichihashi N, Yomo T (2016) Host–parasite oscillation dynamics and evolution in a compartmentalized RNA replication system. Proc Natl Acad Sci 113:4045–4050
82. Elton C, Nicholson M (1942) The ten-year cycle in numbers of the lynx in Canada. J Anim Ecol 11:215–244
83. Uffaker BH (1958) Experimental studies on predation: dispersion factors and predator·prey oscillations. Hilgardia 27(14):343–383
84. Utida S (1957) Cyclic fluctuations of population density intrinsic to the host-parasite system. Ecology 38:442–449
85. Hudson PJ, Dobson AP, Newborn D (1998) Prevention of population cycles by parasite removal. Science 282:2256–2258
86. Mizuuchi R, Furubayashi T, Ichihashi N (2022) Evolutionary transition from a single RNA replicator to a multiple replicator network. Nat Commun 13
87. Yukawa K, Mizuuchi R, Ichihashi N (2022) Relaxed substrate specificity in Qβ replicase through long-term in vitro evolution. Life 12:32
88. Yoshiyama T, Ichii T, Yomo T, Ichihashi N (2018) Automated in vitro evolution of a translation-coupled RNA replication system in a droplet flow reactor. Sci Rep 8:11867
89. Mizuuchi R, Ichihashi N (2018) Sustainable replication and coevolution of cooperative RNAs in an artificial cell-like system. Nat Ecol Evol 2:1654–1660
90. Forster AC, Church GM (2006) Towards synthesis of a minimal cell. Mol Syst Biol 2
91. Sakatani Y, Yomo T, Ichihashi N (2018) Self-replication of circular DNA by a self-encoded DNA polymerase through rolling-circle replication and recombination. Sci Rep 8:13089
92. Okauchi H, Ichihashi N (2021) Continuous cell-free replication and evolution of artificial genomic DNA in a compartmentalized gene expression system. ACS Synth Biol 10:3507–3517
93. Libicher K, Hornberger R, Heymann M, Mutschler H (2020) In vitro self-replication and multicistronic expression of large synthetic genomes. Nat Commun 11:1–8
94. Brookwell A, Oza JP, Caschera F (2021) Biotechnology applications of cell-free expression systems. Life 11:1367
95. Zhang Y, Huang Q, Deng Z, Xu Y, Liu T (2018) Enhancing the efficiency of cell-free protein synthesis system by systematic titration of transcription and translation components. Biochem Eng J 138:47–53
96. Murray CJ, Baliga R (2013) Cell-free translation of peptides and proteins: from high throughput screening to clinical production. Curr Opin Chem Biol 17:420–426
97. Fujiwara K, Katayama T, Nomura S-IM (2013) Cooperative working of bacterial chromosome replication proteins generated by a reconstituted protein expression system. Nucleic Acids Res 41:7176–7183
98. Su'etsugu M, Takada H, Katayama T, Tsujimoto H (2017) Exponential propagation of large circular DNA by reconstitution of a chromosome-replication cycle. Nucleic Acids Res 45: 11525–11534
99. Hagino K, Ichihashi N (2023) In vitro transcription/translation-coupled DNA replication through partial regeneration of 20 Aminoacyl-tRNA Synthetases. ACS Synth Biol 12(4):1252–1263
100. Davidson EA, Meyer AJ, Ellefson JW, Levy M, Ellington AD (2012) An in vitro autogene. ACS Synth Biol 1:190–196
101. Miyachi R, Shimizu Y, Ichihashi N (2022) Transfer RNA synthesis-coupled translation and DNA replication in a reconstituted transcription/translation system. ACS Synth Biol 11:2791–2799
102. Fujii S et al (2014) Liposome display for in vitro selection and evolution of membrane proteins. Nat Protoc 9:1578–1591

Adv Biochem Eng Biotechnol (2023) 186: 141–162
https://doi.org/10.1007/10_2023_223

Published online: 15 June 2023

Rapid and Finely-Tuned Expression for Deployable Sensing Applications

Alexandra T. Patterson and Mark P. Styczynski

Contents

1 Introduction 142
2 Sensing and Output Elements 144
2.1 Sensing Elements 144
2.2 Output Elements 145
3 Cell-Free Biosensor Optimization 148
3.1 Tuning Genetic Circuit Template Concentrations 148
3.2 Optimization of Regulatory Sequences 150
3.3 Chemical Buffer Optimization 150
3.4 Lysate Optimization 151
4 Additional Considerations for Biosensor Deployment 152
4.1 Portability and Stability 152
4.2 Cost-Effectiveness 153
4.3 Robustness to Matrix Effects 154
5 Perspectives 155
6 Conclusion 156
References 157

Abstract Organisms from across the tree of life have evolved highly efficient mechanisms for sensing molecules of interest using biomolecular machinery that can in turn be quite valuable for the development of biosensors. However, purification of such machinery for use in in vitro biosensors is costly, while the use of whole cells as in vivo biosensors often leads to long sensor response times and unacceptable sensitivity to the chemical makeup of the sample. Cell-free expression systems overcome these weaknesses by removing the requirements associated with maintaining living sensor cells, allowing for increased function in toxic environments and rapid sensor readout at a production cost that is often more reasonable

A. T. Patterson and M. P. Styczynski (✉)
School of Chemical and Biomolecular Engineering, Georgia Institute of Technology, Atlanta, GA, USA
e-mail: alexandrapatterson@gatech.edu; mark.styczynski@chbe.gatech.edu

than purification. Here, we focus on the challenge of implementing cell-free protein expression systems that meet the stringent criteria required for them to serve as the basis for field-deployable biosensors. Fine-tuning expression to meet these requirements can be achieved through careful selection of the sensing and output elements, as well as through optimization of reaction conditions via tuning of DNA/RNA concentrations, lysate preparation methods, and buffer conditions. Through careful sensor engineering, cell-free systems can continue to be successfully used for the production of tightly regulated, rapidly expressing genetic circuits for biosensors.

Graphical Abstract

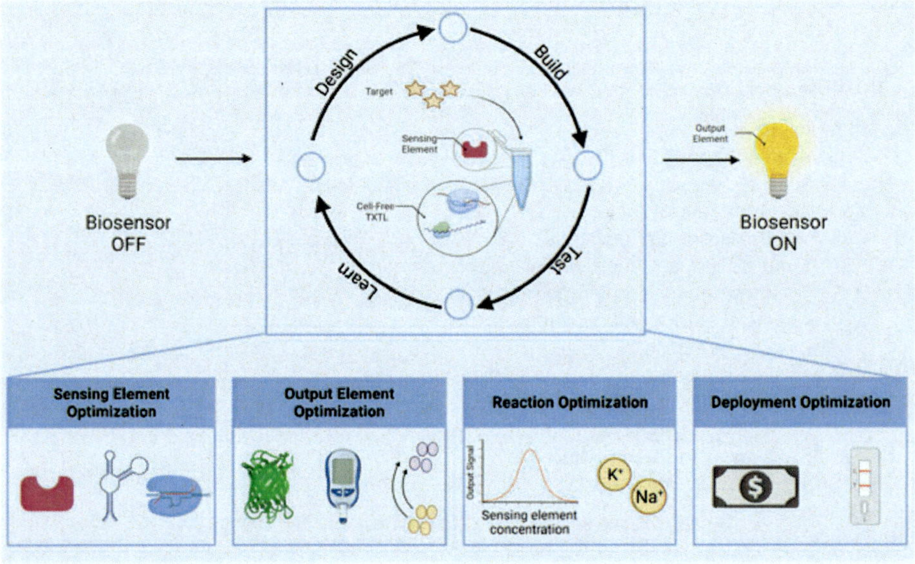

Keywords Biosensors, Cell-free expression, Genetic circuit, Lyophilization, Rapid expression

1 Introduction

Organisms are continually sensing their surroundings and subsequently regulating intracellular levels of RNA and protein to best adapt to their extracellular environment. For example, *Ochrobactrum tritici,* a gram-negative bacterium isolated from wastewater plant sludge containing high levels of chromium, has evolved chromate responsive-transcription factors [1]. Likewise, Pseudomonas sp. strain ADP evolved in soils and water contaminated with the herbicide atrazine and due to the resulting environmental pressure has evolved atrazine-responsive transcription factors

[2]. These contaminant-responsive factors sense the levels of chromium or atrazine in the surroundings and regulate the expression of genes to best adapt to the extracellular environment [1, 2]. Organisms contain a plethora of sensing mechanisms like these not just for contaminant detection, but also for general sensing of the intracellular and extracellular levels of small molecules, proteins, ions, and nucleotides. These mechanisms are a powerful tool that can be adapted into biosensors by rewiring their outputs to express easily interpretable signals, such as a fluorescent protein, in the presence of a target of interest. This makes organisms and their respective sensing elements a promising platform for biosensor development.

Biosensors can be broadly defined as devices or assays that exploit biological sensing mechanisms to generate a measurable signal in the presence of a target. Effective biosensors must be specific (not generate signal due to the presence of other analytes) and sensitive (able to detect often very low concentrations of the target). To be field deployable, they should be portable, user-friendly, and low-cost [3–5], as well as provide results rapidly and without the use of expensive equipment [6–8]. Achieving high specificity and sensitivity within a field-deployable format is a significant challenge that typically requires fine-tuning of the biological sensing machinery.

Traditional approaches to effectively harnessing biological sensing machinery involve either purification or whole cells [5]. In purification-based approaches, the biological machinery (e.g., transcription factors or aptamers) is isolated and purified for use in a device or assay. Although highly sensitive, this approach can be costly and difficult to deploy due to the purification and subsequent cold chain storage requirements of such systems [9]. Whole-cell biosensors address some of these weaknesses by synthesizing the biological sensing machinery in vivo, decreasing the cost. However, these sensors typically must remain alive to maintain function, causing complications due to potential toxicity of chemicals in the test sample as well as due to the output signal. Furthermore, the cell membrane limits transport into the cell, potentially decreasing sensitivity and increasing time to sensor response [3, 10].

Cell-free protein expression systems (CFPS) have the potential to address all of the limitations described above. Specifically, CFPS exhibit rapid protein production for fast response, shelf stability for field deployability, and a membraneless design to avoid transport limitations [11, 12]. The membraneless nature of CFPS even makes the engineering and optimization of sensors more straightforward, as expression levels can be controlled by simply adjusting template concentrations instead of requiring cloning to screen different promoters or ribosomal binding sites (RBS). CFPS can function in a variety of chemical environments including serum, urine, and wastewater and can synthesize toxic proteins at high concentrations [3, 10, 11].

For these reasons, CFPS are an ideal platform for deployable biosensor development. However, the use of CFPS for biosensors requires significant optimization to achieve the rapid and fine-tuned expression required for deployable sensing applications. This chapter will discuss development and optimization strategies for cell-free biosensors including the careful selection of the sensing and output elements and their relative concentrations, as well as optimization of reaction conditions including the lysate and buffer composition. Additionally, methods to achieve field deployment, including lowering cost, increasing portability, and accounting for matrix effects, will be discussed.

2 Sensing and Output Elements

While cell-free biosensors can be implemented in many different ways, they generally include two types of parts: a sensing element and an output element [13]. The sensing element can detect the presence of a target such as DNA, protein, or ions and subsequently control the output element. Output elements translate this sensed information into an interpretable signal, such as a visual color change or an electrochemical signal. Both the sensing element and output element must be carefully selected and tuned in order to achieve proper sensor behavior; in the context of synthetic biology-based sensors, this often entails tight regulation of transcriptional and translational control to enable quick production of interpretable results.

2.1 Sensing Elements

Selecting the sensing element in a cell-free biosensor (CFB) is a key first step in developing a functioning assay. Of most importance is for the sensing element to be specific, sensitive and exhibit low levels of leak (activation when the target is not present). Sensing elements can be composed of DNA, RNA, or protein, each of which may require different resources and time for production within the CFB.

Some examples of protein-based sensing elements are transcription factors and the clustered regularly interspaced short palindromic repeats (CRISPR-Cas) systems. Transcription factors activate or repress the expression of a gene in the presence of the target, often by binding to the promoter region on the output element [14–18]. CRISPR-Cas is a class of proteins that recognizes specific RNA or DNA sequences. Once bound to the cognate RNA or DNA of interest, the CRISPR-Cas system can regulate gene expression through a variety of methods including interference [19] (blocking polymerase binding) or cleavage (degrading oligonucleotides via endonuclease activity) [20–22].

Alternatively, RNA or DNA strands can be used for the detection of proteins, ions, small molecules, or oligonucleotides. For example, aptamers are RNA or DNA oligonucleotides with evolved secondary structure which can tightly bind proteins, ions, or small molecules and can regulate output expression upon target binding via mechanisms such as transcriptional repression or CRISPR-Cas activation [23, 24]. Riboswitches [25–27] are RNA sensor molecules that regulate gene expression directly via structure switching of the coding sequence to terminate transcription or initiate translation. Riboswitches are comprised of two parts: an aptamer region that binds targets, and the expression platform that controls output production. Both the aptamer region and the expression platform must shift structure upon binding of the target, but the interaction between the aptamer and expression platform is not well understood, making the development of new riboswitches extremely challenging [28, 29]. Similarly, toehold switches are RNA with secondary structure that occludes the RBS and thus inhibits transcription when the target is absent; however, toehold switches are much simpler to design and develop compared to riboswitches, as they use only base-pairing interactions and thus their structures are much easier to predict through established techniques and software [30]. However, toehold switch applications are limited to the detection of oligonucleotides. Of note, RNA-based sensing elements do not require translation for in situ production in CFBs, which can result in a faster response time but can also decrease limits of detection due to the lack of signal amplification at the transcriptional level.

2.2 Output Elements

Unlike sensing elements, output elements must be produced or activated during the cell-free reaction, requiring both time and the consumption of reaction resources. Of particular note, though, the cell-free production demands of a CFB are different than that of cell-free metabolic engineering or protein production, as it is typically not necessary to maximize the total amount of the output element produced in a cell-free sensing reaction. Instead, it is only necessary to synthesize enough output to enable reliable interpretation. Therefore, CFBs are often tuned to allow for fast production of the sensing element even at the expense of total titer. These output elements may take multiple forms, including fluorescent, colorimetric, bioluminescent, chemical, or electrical.

Fluorescent proteins are some of the most commonly used output elements in CFBs. There exists a plethora of options for fluorescent outputs including superfolder green fluorescent protein (sfGFP) and mCherry. Some fluorescent proteins have been engineered to allow for fast folding, reducing the time to sensor output [31]. When produced at high titer these fluorescent proteins can even be visible to the naked eye. High levels of fluorescent proteins can be toxic to whole cells [32] but can be made at high titer in CFPS without toxicity concerns [11]. However, high titer requirements for visualization can increase the time to expression and burden the CFPS [33, 34]. Lower levels of fluorescent output can be interpreted and easily quantified using machinery such as a plate reader, but this can limit the deployability of the biosensor to low-resource areas.

Alternative fluorescent output elements beyond fluorescent proteins have also been coupled with CFBs in efforts to decrease the time to output. For example, fluorescent molecules can be pre-synthesized and attached to a quencher that is cleaved from the fluorescent molecule upon exposure to the target, restoring fluorescence. Implementation of such an approach with CRISPR-Cas13 collateral cleavage has allowed for detection of RNA and DNA in as little as of a sensing reaction, though only after input RNA/DNA amplification [20] (Fig. 1a). Similarly,

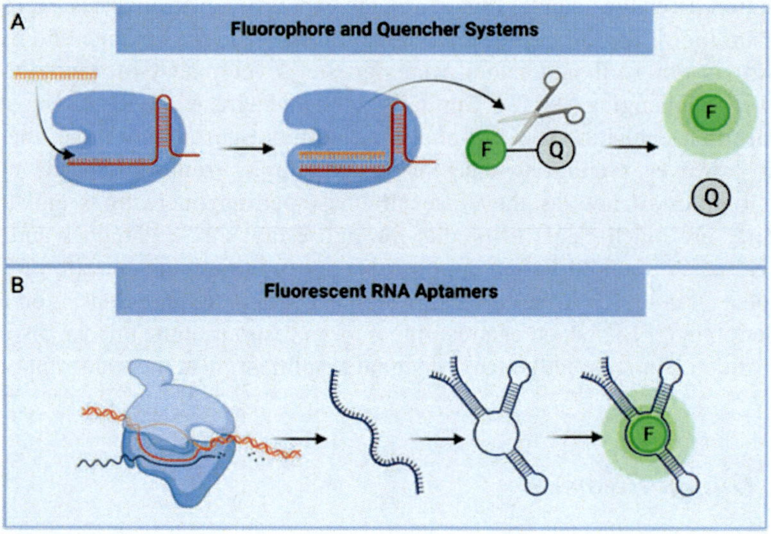

Fig. 1 Approaches to decrease time to sensor output for fluorescent output systems [36]. (**a**) In fluorophore/quencher systems, a quencher (Q) is cleaved from a fluorophore (F) to produce fluorescent signal. For example, one system used activated CRISPR-Cas12/13 can cleave the fluorophore from the quencher on a dual-labeled oligonucleotide reporter in the presence of the target DNA or RNA sequence, resulting in fluorescence [20]. (**b**) When a certain RNA aptamer is transcribed by an RNA polymerase, it folds into a secondary structure that is capable of binding to a small molecule that then becomes fluorescent (F)

fluorescent RNA aptamers have also been used to accelerate the time to sensor response [35]. These aptamers are transcribed during the CFPS in response to the presence of the sensor's target molecule, at which point they bind to small molecules to produce fluorescence (Fig. 1b). 3-way junction dimeric broccoli – a commonly used RNA aptamer – was shown to be twice as fast as sfGFP in producing a detectable signal for a fluoride sensor, likely due to removing the time required for translation [25]. Since translation is not necessary for RNA aptamer-based biosensors, this also opens the door to the use of simpler cell-free lysates or mixtures of recombinant elements at a potential lower cost of manufacture [16].

Colorimetric reporters are advantageous for applications in low-resource areas as they are easily visualized without the use of additional equipment. In CFBs, colorimetric output is achieved via the expression of an enzyme in the presence of the target, which in turn enzymatically converts a colorimetric precursor to yield a visually detectable color change [37]. Enzyme-based colorimetric reporters have been shown to yield faster sensor output and better limits of detection than fluorescent protein reporters, likely due to the signal amplification achieved via enzymatic activity [38]. To date, the most widely-used colorimetric enzyme for CFBs is β-galactosidase (LacZ). When coupled with the substrate chlorophenol red-β-D-galactopyranoside (CPRG), LacZ can create a range of visually distinguishable colors from yellow to purple [37]. However, LacZ is a large protein that can be burdensome and time-consuming to produce; a version of LacZ split into two fragments, LacZα and LacZω, has been used to mitigate this limitation [39]. When the fragments are present by themselves, no enzymatic activity occurs, but when present together the fragments spontaneously reassemble to rescue enzymatic activity. The larger of the two LacZ fragments, LacZω, can be pre-expressed in the lysate and thus present constitutively. Then, when the target is present, only the smaller of the two fragments must be produced to produce a color change. This strategy has been shown to be almost twice as fast in producing a detectable signal compared to expressing the full LacZ protein in a sensing reaction [40]. Beyond LacZ, recent advances have also been made in the use of other colorimetric reporter enzymes in CFBs [37].

Bioluminescent reporters can also potentially be visualized via the naked eye, a key feature for deployable, low-resource biosensors. Bioluminescent reporters produce light via enzymatic reactions and thus require two parts: the enzyme (luciferase) and the substrate (luciferin) [41]. In a cell-free cocaine sensor, firefly luciferase yielded lower LOD than an sfGFP reporter due to its lower signal-to-noise ratio [42]. However, firefly luciferase is a larger protein (62 kDa) [43], requiring additional expression resources and increased synthesis and folding time. To decrease the time to sensor output, the engineered luciferase Nanoluc can be used. Nanoluc is only 19 kDa, has enhanced stability, and has an over 150-fold increase in luminesce compared to two common luciferases, firefly and renilla [41]. It was shown to yield peak signal output in only 7–12 min in a toehold switch COVID biosensor [44] and was separately shown to have equivalent LOD to sfGFP in a shorter time frame [38]. However, the response was short-lived with a quick decay in less than 20 min due to the depletion of the luciferase substrate, which could limit some potential applications. Recent work has moved toward cell-free luciferase systems with substrate regeneration yielding output for 8 h, though the burden of the regeneration system is too large for implementation in a single cell-free reaction [45].

Enzymes that produce other small molecules for non-colorimetric measurements have also been established as cell-free reporters. Specifically, the production of glucose coupled with a glucose monitor has been used for the detection – and in some cases, quantification – of amino acids [46], small molecules [47, 48], and RNA [47, 48]. Via sensing elements like those described above, the presence of a target prompts expression of enzymes such as trehalase [47], invertase [46], or LacZ [48] to convert trehalose, fructose, and lactose (respectively) into glucose. These samples are then loaded into a portable glucose monitor for digital readout. Although these approaches can yield numerical output even in low-resource environments, they often require more user-executed steps and a longer time for sensor output.

Electrical output elements have significant potential due to their high sensitivity. To date, the majority of electrochemical CFBs have relied on functionalized nucleic acids that interact with an electrochemical surface to generate a current [49]. In these systems, a protein such as CRISPR Cas13 [50] or a restriction enzyme [51] is generated or activated in the presence of a target. These biomolecules interact with the functionalized DNA, often cleaving it, resulting in current changes due to the modified interactions between the functionalized DNA and the electrochemical surface. These current changes can be monitored via a potentiostat, allowing for highly sensitive detection. However, the requirement for a potentiostat for readout increases the cost of such a sensor device and decreases its portability and usability in the field [52].

3 Cell-Free Biosensor Optimization

Once the sensing and output elements have been selected, optimization of the sensor can be performed. Optimization of reaction conditions, preparation conditions and protocols, and the sequence elements of the genetic network defining the sensor can decrease the limit of detection, decrease leak, and increase reaction speed, all leading to a more functional sensor. Importantly, the ideal reaction conditions in one sensor design may be different from those in another design meaning that sensor-specific tuning is typically a requirement.

3.1 Tuning Genetic Circuit Template Concentrations

One of the most common and effective CFPS optimization strategies is to control the levels of biomolecular sensing machinery in the reaction, through either control of their expression templates or the upstream steps used to add them directly to the reaction [14, 53–56] (Fig. 2a). The open, membraneless nature of CFPS greatly simplifies tuning of expression template levels compared to whole-cell sensors, as the concentration of DNA/RNA can be adjusted directly rather than via RBS and promoter changes as is required to tune whole-cell systems [11]. This can be particularly critical when using native regulatory sequences to control expression, as changes to those sequences can often lead to deleterious effects. Similarly, sensing machinery can be directly added in protein form after synthesis via other approaches, allowing for even finer tuning of concentrations.

Optimization of template concentrations is particularly important to CFBs because often multiple proteins must be expressed simultaneously in a given reaction, and indirect interactions in the expression of these different proteins can lead to difficult-to-predict results. Expression machinery such as ribosomes and polymerases are present in limited quantities in cell-free reactions, leading to resource competition that affects the relative levels of each expressed protein [33, 34, 57, 58]. This competition can cause genetic circuits to respond in an unpredictable manner when template concentrations are adjusted. These effects can be avoided by working at low DNA template concentrations, thus minimizing competition [33], though this approach can have deleterious impacts on limits of detection, signal output, or time to sensor output. While sequence engineering (described in more detail below) can help overcome some resource limitations [59], other approaches including the expression of proteins prior to the start of the biosensing reaction – as

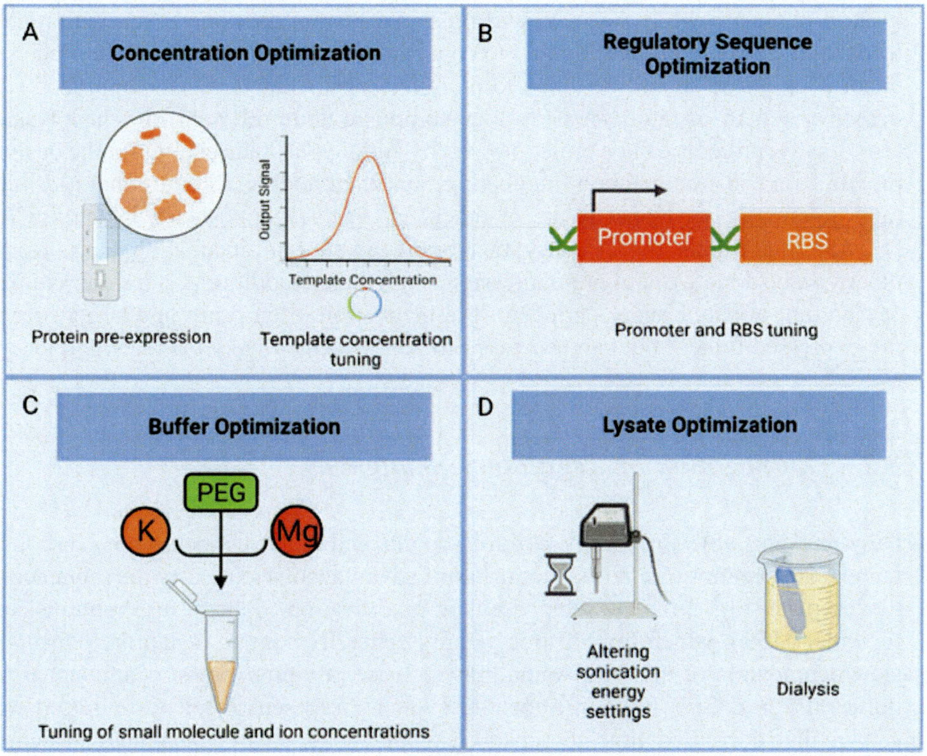

Fig. 2 Methods to optimize CFBs [36]. (**a**) Concentrations of expressed RNA and protein can be optimized via pre-expression of proteins or adjusting template concentrations. (**b**) Concentrations of expressed RNA and protein can be indirectly optimized via promoter and RBS screening, (**c**) Concentrations of small molecules such as magnesium and PEG can be tuned. (**d**) Lysate preparation can be optimized via alteration of sonication and dialysis conditions

opposed to synthesis during the sensing reaction – can also reduce competition for polymerases and ribosomes. It is worth noting, though, that the effects of synthesizing multiple proteins simultaneously can sometimes result in increases in protein output [33, 60] that could benefit sensor functionality. It is hypothesized that these off-target increases may be attributed to "competition" for nucleases, resulting in an increased RNA half-life [61].

Expression of proteins prior to the start of the biosensing reaction can be achieved via purification, lysate enrichment, or pre-incubation. Purification and addition of proteins of interest is the most conceptually simple approach, but it can complicate manufacturing and increase sensor cost [16]. Lysate enrichment is an inexpensive alternative achieved by inducing the expression of a protein of interest before crude extract preparation, yielding a lysate "enriched" with that protein. However, this often results in low-functioning lysate: the cellular burden of unnaturally high levels of (sometime toxic) protein production leads to decreased concentrations or activities of expression machinery. This limitation can be overcome by mixing enriched lysate with unenriched lysate to rescue most expression function yet still with higher levels of the desired protein. Similar to concentration tuning, the ratios of enriched to unenriched lysate should be tuned to optimize sensor function. This approach was demonstrated in an atrazine CFB that combined four different enriched lysates [14]. Lastly, proteins can be pre-expressed by adding plasmids coding for the desired protein to a cell-free reaction, incubating, and then adding all additional plasmids (and potentially additional lysate or other cell-free reagents) needed for the CFB [17, 53]. Importantly, the CFBs can be lyophilized after pre-incubation just as would otherwise be done without pre-expression, such that no additional steps are required at the point of use. Longer periods of pre-incubation often result in a larger titer of pre-expressed protein but can decrease efficiency of the final CFB reaction [17].

3.2 Optimization of Regulatory Sequences

Despite being able to directly control protein and RNA concentrations via their template concentrations, it is sometimes still advantageous to modify the sequence of the template (Fig. 2b). One such case is in modification of the 5′ or 3′ untranslated regions (UTRs), which for cell-free vectors generally contain a stability hairpin to slow degradation of the mRNA and thus increase protein expression and signal-to-noise ratio in a CFB [62]. In a toehold switch-based sensor for norovirus, it was shown that decreased time to sensor readout and decreased leakiness in sensor expression could also be achieved via UTR optimization [63]. Moreover, UTR modifications have been shown to improve protein production for biosensors using both purified recombinant elements (PURE) [63, 64] and crude lysate [62, 65].

Expression regulatory sequence optimization can also be a useful knob for optimizing sensor performance. When using native regulatory elements (e.g., promoters), adjusting the RBS strength can sometimes be critical for achieving detection in the clinically relevant concentration range of the target [59]. Additionally, hybrid promoters comprised of exogenous regulatory elements can be engineered for *E. coli* expression systems by placing a non-native regulatory binding site in front of a native polymerase binding site, allowing for regulated transcription by either *E. coli* or T7 RNA polymerase. It has been shown that optimization of these hybrid promoters and their −35 and −10 binding sites can reduce leak and increase target expression [66].

3.3 Chemical Buffer Optimization

Prototypical cell-free systems include the lysate, buffering salts such as magnesium and potassium, small molecules such as putrescence and spermidine, and expression substrates such as amino acids and nucleotides [12]. The relative concentration of these components has been optimized over the years for reactions whose goals are typically to maximize titer of a single protein [67]. However, these conditions do not necessarily correspond with optimal CFB characteristics and functionality, as finely-tuned production of proteins expressed from complex circuitry is often required; thus, further application-specific optimization of these concentrations in CFBs can help improve these sensors by, for example, minimizing LOD and time to sensor output (Fig. 2c). For example, by simply adjusting the magnesium glutamate concentration, a higher titer of output element (GFP) expressed from a toehold switch-, riboswitch-, and transcription factor-based sensor was achieved, which in a CFB can often be translated into a lower LOD and an increase in the signal-to-noise ratio [25, 59, 68]. Additionally, the concentrations of crowding agents such as PEG and Ficoll have been shown to control the kinetics of a cell-free reaction [69]. The optimal buffer concentration can be a function both of the individual lysate due to batch-to-batch variability [25] and of the sensing components such as transcription factors [59]. This underscores the importance of re-optimizing buffer concentrations for new sensor circuits as well as periodically during manufacturing.

3.4 Lysate Optimization

Selection and optimization of the cell-free preparation protocol is also critical in the development of functional and deployable biosensors. The first major consideration when optimizing the CFB lysate is whether to use crude lysate, which is a non-purified mixture of lysate from whole cells, or PURE a commercially available minimal expression system consisting of reconstituted transcription and translation machinery. PURE may be necessary when expressing from linear DNA templates to avoid degradation due to the presence of exonucleases in crude lysate and can be of great utility during tuning and product development as commercially available PURE currently exhibits less batch-to-batch variability than crude lysate. However, PURE systems are expensive, are generally designed only for transcription from T7 systems, and often have lower protein titers or slower expression than crude extract-based systems [70]. The use of native transcription factors or achievement of higher protein titers may therefore require additional purified components such as alternative polymerases, additional ribosomal proteins and elongation factors, or additional crowding agents such as glycerol [71, 72]. On the other hand, crude extracts exhibit poorly-characterized residual metabolic activity, nuclease activity, and high levels of batch-to-batch variability, all of which affect cell-free reaction dynamics and complicate the implementation of CFBs. Nonetheless, lysate-based systems are significantly cheaper than PURE, an important consideration for CFBs that must be distributed at low cost.

Since crude lysates are often produced in-house, there are more opportunities for optimization of the lysate itself. Crude extracts prepared with a run-off reaction and dialysis have been shown to have significantly larger dynamic ranges for reactions using native transcriptional machinery and sensing circuits based on transcription factors, riboswitches, and toehold switches [65]. The strain of cells used to prepare the crude extract has also been shown to affect CFB performance by up to 400% [73]. Additionally, the use of gentle lysis preparation methods such as low-energy sonication can decrease the limit of detection, as previously seen in an ascorbate sensor, likely due to preservation of membrane-like-structures [15] (Fig. 2d).

4 Additional Considerations for Biosensor Deployment

Once the sensing and output elements have been selected, the physical implementation of the device or assay must also be established to enable field deployment by maximizing portability and stability, minimizing cost, and ensuring biosensor performance in complex matrices. However, changes to the physical implementation of the cell-free sensor may lead to changes in protein expression, requiring interdependent cycles of sensor optimization and implementation design.

4.1 Portability and Stability

Portability and stability at ambient conditions are critical to increasing the field deployability of biosensors at the point of need [4]. Not only do low-resource areas with the greatest needs for sensing and detection often have minimal access to sophisticated instrumentation and equipment, they also often do not have reliable access to electricity, prohibiting the use of sensors that must be maintained at cold temperatures before use (known as requiring a "cold chain") [74]. Even if electricity is accessible, cold chain storage still requires expensive shipping and storage equipment that is beyond the budgets of many of the potential users in these remote areas. CFBs are advantageous for such low-resource settings because they can be lyophilized ("freeze-dried") and stored at room temperature with almost-instant reactivation for use and minimal impact on their function [11, 75] (in contrast to whole-cell sensors that can lose significant activity after lyophilization and require longer lag times for the cells to reactivate their functions).

Although the components of a cell-free reaction – for example, cell lysate, mixtures of small molecules and ions, and DNA/RNA for programming a genetic circuit – can be lyophilized individually and combined immediately before use, lyophilizing all components together allows for a simpler end user experience [76–78]. Lyophilization of all components for a purified CFB has been shown to result in the same kinetics of GFP expression as fresh extracts and has been shown to be stable for over a year at room temperature [78]. Stabilizers such as trehalose and polyethylene glycol (PEG) can be added to the reaction mixture prior to lyophilization to provide additional stabilization of function [76].

CFBs can be lyophilized onto porous matrices such as paper to increase portability. Prior to the addition of reaction components, materials should have blocking agents added to reduce nonspecific interactions between reaction components and the matrix, which can otherwise reduce reaction efficiency [26, 78]. Widely-used blocking agents for molecular biology such as bovine serum albumin and PEG have been shown to increase the performance of GFP expression reactions on paper discs [78]. The type of paper used has been shown to affect protein yield, with blocked cellulose nitrate and nylon decreasing output in a LacZ reaction while cellulose acetate, chromatography paper, and ashless filter paper allowing for strong LacZ synthesis resulting in an easily interpretable visible output [44]. Cell-free extracts have also been incorporated into wearable materials such as woven fabrics and single cotton threads. As in paper-based sensors, the type of wearable material the CFB is embedded in can affect expression yield and time to sensor output [26].

4.2 Cost-Effectiveness

In addition to portability and stability, cell-free extracts must be inexpensive for large-scale deployment to be feasible [4]. While standard diagnostic tests such as polymerase chain reaction and immunoassays can be costly and require expensive equipment and trained personnel, cell-free reactions can be executed simply, without complex instrumentation, and at a reagent-only cost of 35–65 cents per μl for purified recombinant (PURE) reactions or as little as 1.6 cents per μl of lysate-based reaction [78, 79]. These costs could be decreased even further by engineering cell-free systems that use less expensive secondary energy sources for ATP regeneration – the primary source of reaction energy – than the most commonly used options, which are phosphoenolpyruvate (PEP) or 3-phosphoglycerate (3-PGA) [80, 81]. Less expensive sources such as maltodextrin have been used to decrease cost and ultimately increase total protein yield compared to typical energy sources via recycling of inorganic phosphates, an inhibitor of protein synthesis [82–84]; this approach has been successfully implemented in lyophilized RNA sensors [83]. Glucose and glucose-6-phosphate-based energy systems have also been shown to be functional at a fraction of the cost of traditional secondary energy sources, though their use can destabilize the pH of the cell-free reaction [85].

Likewise, advances have been made in decreasing the cost of crude extract preparation and lyophilization. Crude extract preparation, while cheaper than using purified recombinant elements, is time-intensive, labor-intensive, and potentially difficult to scale up [86]. To address these limitations, autolysis strains have been developed that merge the lysis and lyophilization steps by expressing endolysis genes that become active upon disruption of the inner membrane via lyophilization. Promisingly, this system was previously used in a functional mercury sensor, though the autolyzed sensor had a weaker and slower response compared to more widely-used cell-free lysate preparations [86]. Similarly, lyophilization is also challenging to scale up and often requires costly equipment [75]. To try to reduce the cost associated with the lyophilization step of sensor preparation, a low-vacuum room temperature approach has been developed that results in up to 60% recovery of lysate function after 2 weeks. While this is less recovery than from lyophilization, the 60% recovery is likely still sufficient for use in most CFBs [81].

4.3 Robustness to Matrix Effects

CFBs are often used to measure concentrations in complex biological sample matrices such as blood, urine, and saliva. The small molecules, proteins, and ions within these biological matrices often alter expression and function in CFBs, a phenomenon referred to as matrix effects. Overcoming this challenge often requires the use of additional tuning to ensure performance in the expected sample matrix [87]. Matrix effects may be generalizable between samples in a given sample type, [88] such as the presence of proteases and RNases, but also may be sample-specific [16, 54, 87] based on the levels of ions and small molecules in any individual sample.

To address the generalizable matrix effects due to proteases and RNases, inhibitors can be supplemented in the sensor reaction [88] (Fig. 3). This is typically achieved by adding commercially produced inhibitors stored in glycerol. However,

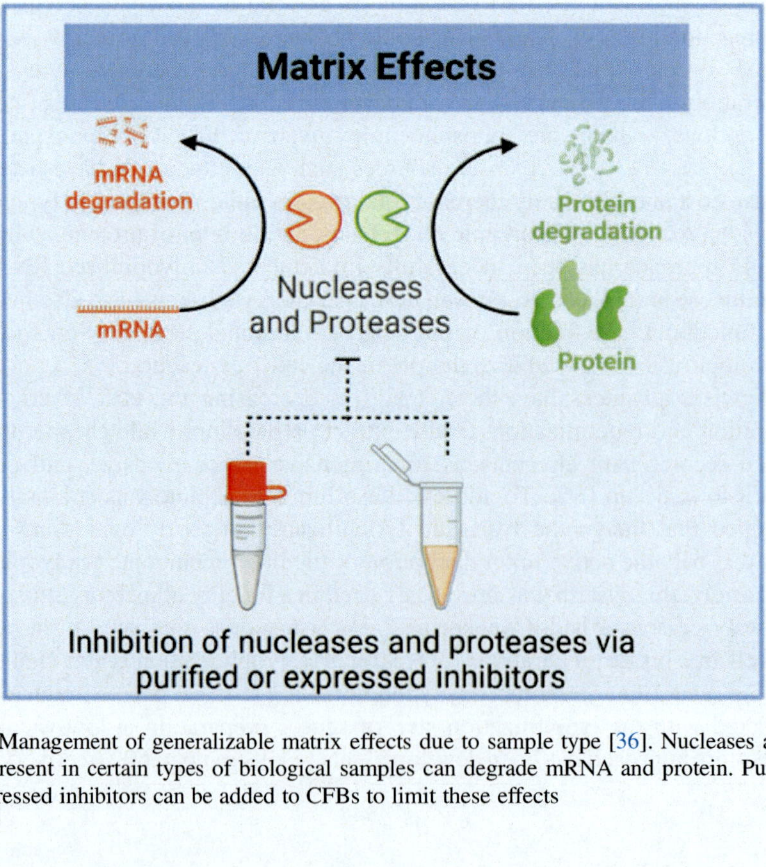

Fig. 3 Management of generalizable matrix effects due to sample type [36]. Nucleases and proteases present in certain types of biological samples can degrade mRNA and protein. Purified or pre-expressed inhibitors can be added to CFBs to limit these effects

glycerol has been shown to inhibit CFBs, mitigating some of the positive effects of inhibitor addition [88]. Furthermore, the addition of a commercial RNAase inhibitor can increase reaction costs by 900% [89]. To address this, recent work has demonstrated that production of RNAse inhibitors via pre-incubation [90] and lysate enrichment [89] is a feasible supplementation approach. Both methods produced enough RNase inhibitor to maintain performance in the presence of biological matrices, at a fraction of the cost of adding purified inhibitor.

Addressing sample-specific matrix effects due to different concentrations of small molecules and ions requires different approaches. One way to address this challenge is to remove specific ions that are of concern via enzymatic activity [91] or chemical reactions [92, 93]. However, these approaches only address individual ions and do not scale well to account for matrix effects as a whole. Another approach beyond engineering the CFB itself is to use parallel calibration. For this method, the same biological sample is added to multiple individual reactions with varying amounts of a limiting reagent, generating custom sample-specific curves that correspond to levels of target corrected for the sample's matrix effects, ultimately allowing for point-of-care quantification [27, 87].

5 Perspectives

CFB optimization is a time- and labor-intensive process due to the multitude of variables that can be manipulated to improve performance. Expediting biosensor optimization and development could decrease costs and allow for quick deployment of CFBs for novel, time-sensitive targets (e.g., a newly-discovered pathogen during a pandemic). Future work should focus on expediting the design, build, test, and learn cycle for CFBs via the development of more accurate computational models of cell-free expression, decreasing batch-to-batch variability of cell-free lysates, expanding possible output elements, and optimizing the speed, lifetime, and titer of cell-free reactions.

Although computational models of CFPS exist for both crude lysate and PURE systems [94–96], they do not capture all the phenomena necessary for true predictive power and thus facile in silico optimization of CFBs. As a result, they have yet to be widely adopted as a preliminary step in CFB design and optimization. Future computational models should interface with models on sensing elements, such as toehold switch design software [30], and account for the impacts of the cell-free chemical environment on sensing element function [97]. Models should be updated to include off-target effects such as resource competition, predict the effects of the pre-expression of sensing proteins, and be accurate at the lower levels of protein production common to CFBs. All of these are key modeling specifications needed to accurately predict leak and signal-to-noise ratios in CFBs, which are critical in determining sensor function and utility. Ideally, prediction of buffer, lysate, and matrix effects would be incorporated as well. Ultimately this would lead to a biosensor design software in which the user inputs a target of interest, and the ideal output and sensing elements are determined with information on optimal concentration, buffer, and lysate characteristics, streamlining experimental tuning.

Importantly, models cannot be widely utilized for crude lysate systems until batch-to-batch variability in lysate preparation is circumvented or controlled for. Current approaches to mitigate batch-to-batch variability at the level of the complete CFB may require re-optimizing genetic circuit template concentrations and/or chemical buffers with every new batch of lysate to yield consistent sensor output, further increasing the time and effort required for biosensor development. Furthermore, such a continuous optimization approach presents major challenges to industrialized, commercial production of crude lysate-based CFBs.

Future work should also focus on expanding and optimizing the repertoire of output elements that can be used in CFBs, with a focus on options that require minimal equipment for readability. For example, recent work expanded the luminescent output elements that can be used in CFBs from two to four [45]. Having a panel of output elements which do not require equipment for interpretation would allow for multiplexing – the testing of multiple substrates in the same reaction [37] – at the point-of-care and would increase the number of "knobs" that can be tuned during CFB development to reach specific development goals. Furthermore, work should be done to engineer more efficient output elements, similar to a recently developed brighter fluorescent protein that may be detectable at lower limits of detection [98], or outputs that can be synthesized quickly in a cell-free reaction with minimal burden on the expression machinery.

Finally, CFBs are often limited by the cell-free lysate itself. Generating more efficient lysate that can produce protein faster or at higher levels would enable more sensitive and rapidly-responding CFBs. Specifically, extending the lifetime of batch cell-free reactions could enable long-term monitoring using CFBs, while generating lysate with the potential for high protein titers could minimize the impacts of resource competition, eliminate the need for pre-expression, and simplify design and optimization.

6 Conclusion

Cell-free reactions are an optimal platform for biosensor development due to their shelf stability, speed, and robustness to toxic conditions and environments. CFBs are a unique application of cell-free reactions, as they require finely-tuned titers of sensing and output elements to reliably detect targets, which in turn often requires significant optimization to achieve. To date, successful CFBs have been developed for the detection of a wide variety of targets including oligonucleotides, small molecules, proteins, and more. Expanding our understanding of complex genetic circuits, improving our ability to predict the output of cell-free reactions, and enhancing expression efficiency would all open new doors for the field of CFBs.

Acknowledgments This work was supported by the National Institutes of Health (R01-EB022592) and the National Science Foundation (DGE-2039655).

References

1. Branco R, Morais PV (2013) Identification and characterization of the transcriptional regulator ChrB in the chromate resistance determinant of Ochrobactrum tritici 5bvl1. PloS One 8:e77987. https://doi.org/10.1371/journal.pone.0077987
2. Govantes F, García-González V, Porrúa O et al (2010) Regulation of the atrazine-degradative genes in pseudomonas sp. strain ADP: regulation of atrazine biodegradation. FEMS Microbiol Lett 310:1–8. https://doi.org/10.1111/j.1574-6968.2010.01991.x
3. Brooks SM, Alper HS (2021) Applications, challenges, and needs for employing synthetic biology beyond the lab. Nat Commun 12:1390. https://doi.org/10.1038/s41467-021-21740-0
4. Land KJ, Boeras DI, Chen X-S et al (2019) REASSURED diagnostics to inform disease control strategies, strengthen health systems and improve patient outcomes. Nat Microbiol 4:46–54. https://doi.org/10.1038/s41564-018-0295-3
5. Pancrazio JJ, Whelan JP, Borkholder DA et al (1999) Development and application of cell-based biosensors. Ann Biomed Eng 27:697–711. https://doi.org/10.1114/1.225
6. Mitchler MM, Garcia JM, Montero NE, Williams GJ (2021) Transcription factor-based biosensors: a molecular-guided approach for natural product engineering. Curr Opin Biotechnol 69:172–181. https://doi.org/10.1016/j.copbio.2021.01.008
7. Hicks M, Bachmann TT, Wang B (2020) Synthetic biology enables programmable cell-based biosensors. ChemPhysChem 21:132–144. https://doi.org/10.1002/cphc.201900739
8. Zhang J, Pang Q, Wang Q et al (2022) Modular tuning engineering and versatile applications of genetically encoded biosensors. Crit Rev Biotechnol 42:1010–1027. https://doi.org/10.1080/07388551.2021.1982858
9. Gui Q, Lawson T, Shan S et al (2017) The application of whole cell-based biosensors for use in environmental analysis and in medical diagnostics. Sensors 17:1623. https://doi.org/10.3390/s17071623
10. Tinafar A, Zhou Y, Hong F et al (2022) Cell-Free biosensors: synthetic biology without Borders. In: Thouand G (ed) Handbook of cell biosensors. Springer, Cham, pp 243–281
11. Silverman AD, Karim AS, Jewett MC (2020) Cell-free gene expression: an expanded repertoire of applications. Nat Rev Genet 21:151–170. https://doi.org/10.1038/s41576-019-0186-3
12. Garenne D, Haines MC, Romantseva EF et al (2021) Cell-free gene expression. Nat Rev Methods Primer 1:1–18. https://doi.org/10.1038/s43586-021-00046-x
13. Voyvodic PL, Bonnet J (2020) Cell-free biosensors for biomedical applications. Curr Opin Biomed Eng 13:9–15. https://doi.org/10.1016/j.cobme.2019.08.005
14. Silverman AD, Akova U, Alam KK et al (2020) Design and optimization of a cell-free atrazine biosensor. ACS Synth Biol 9:671–677. https://doi.org/10.1021/acssynbio.9b00388
15. Piorino F, Styczynski MP (2022) Harnessing Escherichia coli's native machinery for detection of vitamin C (ascorbate) deficiency. ACS Synth Biol 11:3592. https://doi.org/10.1021/acssynbio.2c00335
16. Jung JK, Alam KK, Lucks JB (2022) ROSALIND: rapid detection of chemical contaminants with in vitro transcription in vitro transcription factor-based biosensors biosensors. In: Karim AS, Jewett MC (eds) Cell-free gene expression: methods and protocols. Springer, New York, pp 325–342
17. Pandi A, Grigoras I, Borkowski O, Faulon J-L (2019) Optimizing cell-free biosensors to monitor enzymatic production. ACS Synth Biol 8:1952–1957. https://doi.org/10.1021/acssynbio.9b00160
18. Jung JK, Archuleta CM, Alam KK, Lucks JB (2022) Programming cell-free biosensors with DNA strand displacement circuits. Nat Chem Biol 18:385–393. https://doi.org/10.1038/s41589-021-00962-9
19. Marshall R, Maxwell CS, Collins SP et al (2018) Rapid and scalable characterization of CRISPR technologies using an E. coli cell-free transcription-translation system. Mol Cell 69: 146–157.e3. https://doi.org/10.1016/j.molcel.2017.12.007

20. Kellner MJ, Koob JG, Gootenberg JS et al (2019) SHERLOCK: nucleic acid detection with CRISPR nucleases. Nat Protoc 14:2986–3012. https://doi.org/10.1038/s41596-019-0210-2
21. Mahas A, Wang Q, Marsic T, Mahfouz MM (2022) Development of Cas12a-based cell-free small-molecule biosensors via allosteric regulation of CRISPR Array expression. Anal Chem 94:4617–4626. https://doi.org/10.1021/acs.analchem.1c04332
22. Jiao C, Sharma S, Dugar G et al (2021) Noncanonical crRNAs derived from host transcripts enable multiplexable RNA detection by Cas9. Science 372:941–948. https://doi.org/10.1126/science.abe7106
23. Iyer S, Doktycz MJ (2014) Thrombin-mediated transcriptional regulation using DNA aptamers in DNA-based cell-free protein synthesis. ACS Synth Biol 3:340–346. https://doi.org/10.1021/sb4000756
24. Chen J, Zhuang X, Zheng J et al (2021) Aptamer-based cell-free detection system to detect target protein. Synth Syst Biotechnol 6:209–215. https://doi.org/10.1016/j.synbio.2021.07.004
25. Thavarajah W, Silverman AD, Verosloff MS et al (2020) Point-of-use detection of environmental fluoride via a cell-free riboswitch-based biosensor. ACS Synth Biol 9:10–18. https://doi.org/10.1021/acssynbio.9b00347
26. Nguyen PQ, Soenksen LR, Donghia NM et al (2021) Wearable materials with embedded synthetic biology sensors for biomolecule detection. Nat Biotechnol 39:1366–1374. https://doi.org/10.1038/s41587-021-00950-3
27. Harbaugh SV, Silverman AD, Chushak YG et al (2022) Engineering a synthetic dopamine-responsive riboswitch for in vitro biosensing. ACS Synth Biol 11:2275–2283. https://doi.org/10.1021/acssynbio.1c00560
28. Chau THT, Mai DHA, Pham DN et al (2020) Developments of riboswitches and toehold switches for molecular detection-biosensing and molecular diagnostics. Int J Mol Sci 21: E3192. https://doi.org/10.3390/ijms21093192
29. Synthetic riboswitches: from plug and pray toward plug and play. Biochemistry. https://pubs.acs.org. https://doi.org/10.1021/acs.biochem.6b01218. Accessed 2 Dec 2022
30. Zadeh JN, Steenberg CD, Bois JS et al (2011) NUPACK: analysis and design of nucleic acid systems. J Comput Chem 32:170–173. https://doi.org/10.1002/jcc.21596
31. Pédelacq J-D, Cabantous S, Tran T et al (2006) Engineering and characterization of a superfolder green fluorescent protein. Nat Biotechnol 24:79–88. https://doi.org/10.1038/nbt1172
32. Oscar TP, Dulal K, Boucaud D (2006) Transformation of Escherichia coli K-12 with a high-copy plasmid encoding the green fluorescent protein reduces growth: implications for predictive microbiology. J Food Prot 69:276–281. https://doi.org/10.4315/0362-028x-69.2.276
33. Siegal-Gaskins D, Tuza ZA, Kim J et al (2014) Gene circuit performance characterization and resource usage in a cell-free "Breadboard". ACS Synth Biol 3:416–425. https://doi.org/10.1021/sb400203p
34. Borkowski O, Bricio C, Murgiano M et al (2018) Cell-free prediction of protein expression costs for growing cells. Nat Commun 9:1457. https://doi.org/10.1038/s41467-018-03970-x
35. Nutiu R, Li Y (2005) Aptamers with fluorescence-signaling properties. Methods 37:16–25. https://doi.org/10.1016/j.ymeth.2005.07.001
36. Created with BioRender.com
37. Sharpes CE, McManus JB, Blum SM et al (2021) Assessment of colorimetric reporter enzymes in the PURE system. ACS Synth Biol 10:3205–3208. https://doi.org/10.1021/acssynbio.1c00360
38. Lopreside A, Wan X, Michelini E et al (2019) Comprehensive profiling of diverse genetic reporters with application to whole-cell and cell-free biosensors. Anal Chem 91:15284–15292. https://doi.org/10.1021/acs.analchem.9b04444
39. Nath A, Mondal S, Chakraborty S et al (2014) Production, purification, characterization, immobilization, and application of β-galactosidase: a review. Asia Pac J Chem Eng 9:330–348. https://doi.org/10.1002/apj.1801

40. Ma D, Shen L, Wu K et al (2018) Low-cost detection of norovirus using paper-based cell-free systems and synbody-based viral enrichment. Synth Biol 3:ysy018. https://doi.org/10.1093/synbio/ysy018
41. England CG, Ehlerding EB, Cai W (2016) NanoLuc: a small luciferase is brightening up the field of bioluminescence. Bioconjug Chem 27:1175–1187. https://doi.org/10.1021/acs.bioconjchem.6b00112
42. Voyvodic PL, Pandi A, Koch M et al (2019) Plug-and-play metabolic transducers expand the chemical detection space of cell-free biosensors. Nat Commun 10:1697. https://doi.org/10.1038/s41467-019-09722-9
43. Thorne N, Inglese J, Auld DS (2010) Illuminating insights into firefly luciferase and other bioluminescent reporters used in chemical biology. Chem Biol 17:646–657. https://doi.org/10.1016/j.chembiol.2010.05.012
44. Hunt JP, Zhao EL, Free TJ et al (2022) Towards detection of SARS-CoV-2 RNA in human saliva: a paper-based cell-free toehold switch biosensor with a visual bioluminescent output. N Biotechnol 66:53–60. https://doi.org/10.1016/j.nbt.2021.09.002
45. Sato W, Rasmussen M, Deich C et al (2022) Expanding luciferase reporter systems for cell-free protein expression. Sci Rep 12:11489. https://doi.org/10.1038/s41598-022-15624-6
46. Jang Y-J, Lee K-H, Yoo TH, Kim D-M (2019) Interfacing a personal glucose meter with cell-free protein synthesis for rapid analysis of amino acids. Anal Chem 91:2531–2535. https://doi.org/10.1021/acs.analchem.8b05526
47. Amalfitano E, Karlikow M, Norouzi M et al (2021) A glucose meter interface for point-of-care gene circuit-based diagnostics. Nat Commun 12:724. https://doi.org/10.1038/s41467-020-20639-6
48. Zhang Y, Steppe PL, Kazman MW, Styczynski MP (2021) Point-of-care analyte quantification and digital readout via lysate-based cell-free biosensors interfaced with personal glucose monitors. ACS Synth Biol 10:2862–2869. https://doi.org/10.1021/acssynbio.1c00282
49. Yu P, Lei C, Nie Z (2022) Integration of electrochemical interface and cell-free synthetic biology for biosensing. J Electroanal Chem 911:116209. https://doi.org/10.1016/j.jelechem.2022.116209
50. Xu W, Jin T, Dai Y, Liu CC (2020) Surpassing the detection limit and accuracy of the electrochemical DNA sensor through the application of CRISPR Cas systems. Biosens Bioelectron 155:112100. https://doi.org/10.1016/j.bios.2020.112100
51. Sadat Mousavi P, Smith SJ, Chen JB et al (2020) A multiplexed, electrochemical interface for gene-circuit-based sensors. Nat Chem 12:48–55. https://doi.org/10.1038/s41557-019-0366-y
52. Campuzano S, Pedrero M, Yáñez-Sedeño P, Pingarrón JM (2021) New challenges in point of care electrochemical detection of clinical biomarkers. Sens Actuators B 345:130349. https://doi.org/10.1016/j.snb.2021.130349
53. Beabout K, Bernhards CB, Thakur M et al (2021) Optimization of heavy metal sensors based on transcription factors and cell-free expression systems. ACS Synth Biol 10:3040–3054. https://doi.org/10.1021/acssynbio.1c00331
54. Soudier P, Zúñiga A, Duigou T et al (2022) PeroxiHUB: a modular cell-free biosensing platform using H2O2 as signal integrator. ACS Synth Biol 11:2578–2588. https://doi.org/10.1021/acssynbio.2c00138
55. Soudier P, Faure L, Kushwaha M, Faulon J-L (2022) Cell-free biosensors biosensors and AI artificial intelligence (AI) integration. In: Karim AS, Jewett MC (eds) Cell-free gene expression: methods and protocols. Springer, New York, pp 303–323
56. Zhang Y, Zhao C, Bi H et al (2022) A cell-free paper-based biosensor dependent on allosteric transcription factors (aTFs) for on-site detection of harmful metals Hg2+ and Pb2+ in water. J Hazard Mater 438:129499. https://doi.org/10.1016/j.jhazmat.2022.129499
57. Gyorgy A, Murray RM (2016) Quantifying resource competition and its effects in the TX-TL system. In: 2016 IEEE 55th conference on decision and control (CDC). pp 3363–3368
58. Halter W, Allgower F, Murray RM, Gyorgy A (2018) Optimal experiment design and leveraging competition for shared resources in cell-free extracts. In: 2018 IEEE conference on decision and control (CDC). pp 1872–1879

59. Zhang P, Feng H, Yang J et al (2019) Detection of inorganic ions and organic molecules with cell-free biosensing systems. J Biotechnol 300:78–86. https://doi.org/10.1016/j.jbiotec.2019.05.011
60. Miguez AM, Zhang Y, Piorino F, Styczynski MP (2021) Metabolic dynamics in *Escherichia coli* -based cell-free systems. ACS Synth Biol 10:2252–2265. https://doi.org/10.1021/acssynbio.1c00167
61. McSweeney MA, Styczynski MP (2021) Effective use of linear DNA in cell-free expression systems. Front Bioeng Biotechnol 9:715328. https://doi.org/10.3389/fbioe.2021.715328
62. Gräwe A, Dreyer A, Vornholt T et al (2019) A paper-based, cell-free biosensor system for the detection of heavy metals and date rape drugs. PloS One 14:e0210940. https://doi.org/10.1371/journal.pone.0210940
63. Zhang Q, Ma D, Wu F et al (2021) Predictable control of RNA lifetime using engineered degradation-tuning RNAs. Nat Chem Biol 17:828–836. https://doi.org/10.1038/s41589-021-00816-4
64. Chizzolini F, Forlin M, Yeh Martín N et al (2017) Cell-free translation is more variable than transcription. ACS Synth Biol 6:638–647. https://doi.org/10.1021/acssynbio.6b00250
65. Silverman AD, Kelley-Loughnane N, Lucks JB, Jewett MC (2019) Deconstructing cell-free extract preparation for in vitro activation of transcriptional genetic circuitry. ACS Synth Biol 8: 403–414. https://doi.org/10.1021/acssynbio.8b00430
66. Liu X, Silverman AD, Alam KK et al (2020) Design of a transcriptional biosensor for the portable, on-demand detection of cyanuric acid. ACS Synth Biol 9:84–94. https://doi.org/10.1021/acssynbio.9b00348
67. Shin J, Noireaux V (2010) Efficient cell-free expression with the endogenous E. Coli RNA polymerase and sigma factor 70. J Biol Eng 4:8. https://doi.org/10.1186/1754-1611-4-8
68. Batista AC, Levrier A, Soudier P et al (2022) Differentially optimized cell-free buffer enables robust expression from unprotected linear DNA in exonuclease-deficient extracts. ACS Synth Biol 11:732–746. https://doi.org/10.1021/acssynbio.1c00448
69. Vezeau GE, Salis HM (2021) Tuning cell-free composition controls the time delay, dynamics, and productivity of TX-TL expression. ACS Synth Biol 10:2508–2519. https://doi.org/10.1021/acssynbio.1c00136
70. Hillebrecht JR, Chong S (2008) A comparative study of protein synthesis in in vitro systems: from the prokaryotic reconstituted to the eukaryotic extract-based. BMC Biotechnol 8:58. https://doi.org/10.1186/1472-6750-8-58
71. Li J, Gu L, Aach J, Church GM (2014) Improved cell-free RNA and protein synthesis system. PloS One 9:e106232. https://doi.org/10.1371/journal.pone.0106232
72. Sheahan T, Wieden H-J (2022) Ribosomal protein S1 improves the protein yield of an in vitro reconstituted cell-free translation system. ACS Synth Biol 11:1004–1008. https://doi.org/10.1021/acssynbio.1c00514
73. Wen KY, Cameron L, Chappell J et al (2017) A cell-free biosensor for detecting quorum sensing molecules in P. aeruginosa-infected respiratory samples. ACS Synth Biol 6:2293–2301. https://doi.org/10.1021/acssynbio.7b00219
74. Malekjahani A, Sindhwani S, Syed AM, Chan WCW (2019) Engineering steps for mobile point-of-care diagnostic devices. Acc Chem Res 52:2406–2414. https://doi.org/10.1021/acs.accounts.9b00200
75. Pardee K (2018) Perspective: solidifying the impact of cell-free synthetic biology through lyophilization. Biochem Eng J 138:91–97. https://doi.org/10.1016/j.bej.2018.07.008
76. Karig DK, Bessling S, Thielen P et al (2017) Preservation of protein expression systems at elevated temperatures for portable therapeutic production. J R Soc Interface 14:20161039. https://doi.org/10.1098/rsif.2016.1039
77. Smith MT, Berkheimer SD, Werner CJ, Bundy BC (2014) Lyophilized Escherichia coli-based cell-free systems for robust, high-density, long-term storage. Biotechniques 56:186–193. https://doi.org/10.2144/000114158

78. Pardee K, Green AA, Ferrante T et al (2014) Paper-based synthetic gene networks. Cell 159: 940–954. https://doi.org/10.1016/j.cell.2014.10.004
79. Biomolecular breadboards: protocols: cost estimate – OpenWetWare. https://openwetware.org/wiki/Biomolecular_Breadboards:Protocols:cost_estimate. Accessed 1 Nov 2022
80. Kim H-C, Kim D-M (2009) Methods for energizing cell-free protein synthesis. J Biosci Bioeng 108:1–4. https://doi.org/10.1016/j.jbiosc.2009.02.007
81. Guzman-Chavez F, Arce A, Adhikari A et al (2022) Constructing cell-free expression systems for low-cost access. ACS Synth Biol 11:1114–1128. https://doi.org/10.1021/acssynbio.1c00342
82. Caschera F, Noireaux V (2015) A cost-effective polyphosphate-based metabolism fuels an all E. coli cell-free expression system. Metab Eng 27:29–37. https://doi.org/10.1016/j.ymben.2014.10.007
83. Arce A, Guzman Chavez F, Gandini C et al (2021) Decentralizing cell-free RNA sensing with the use of low-cost cell extracts. Front Bioeng Biotechnol 9:727584. https://doi.org/10.3389/fbioe.2021.727584
84. Wang Y, Zhang Y-HP (2009) Cell-free protein synthesis energized by slowly-metabolized maltodextrin. BMC Biotechnol 9:58. https://doi.org/10.1186/1472-6750-9-58
85. Calhoun KA, Swartz JR (2005) Energizing cell-free protein synthesis with glucose metabolism. Biotechnol Bioeng 90:606–613. https://doi.org/10.1002/bit.20449
86. Didovyk A, Tonooka T, Tsimring L, Hasty J (2017) Rapid and scalable preparation of bacterial lysates for cell-free gene expression. ACS Synth Biol 6:2198–2208. https://doi.org/10.1021/acssynbio.7b00253
87. McNerney MP, Zhang Y, Steppe P et al (2019) Point-of-care biomarker quantification enabled by sample-specific calibration. Sci Adv 5:eaax4473. https://doi.org/10.1126/sciadv.aax4473
88. Voyvodic PL, Conejero I, Mesmoudi K et al (2022) Evaluating and mitigating clinical samples matrix effects on TX-TL cell-free performance. Sci Rep 12:13785. https://doi.org/10.1038/s41598-022-17583-4
89. Soltani M, Bundy BC (2022) Streamlining cell-free protein synthesis biosensors for use in human fluids: in situ RNase inhibitor production during extract preparation. Biochem Eng J 177:108158. https://doi.org/10.1016/j.bej.2021.108158
90. Soltani M, Hunt JP, Bundy BC (2021) Rapid RNase inhibitor production to enable low-cost, on-demand cell-free protein synthesis biosensor use in human body fluids. Biotechnol Bioeng 118:3973–3983. https://doi.org/10.1002/bit.27874
91. Zhang J, Xiang Y, Wang M et al (2016) Dose-dependent response of personal glucose meters to nicotinamide coenzymes: applications to point-of-care diagnostics of many non-glucose targets in a single step. Angew Chem Int Ed Engl 55:732–736. https://doi.org/10.1002/anie.201507563
92. Gupta S, Sarkar S, Katranidis A, Bhattacharya J (2019) Development of a cell-free optical biosensor for detection of a broad range of mercury contaminants in water: a plasmid DNA-based approach. ACS Omega 4:9480–9487. https://doi.org/10.1021/acsomega.9b00205
93. Jung JK, Alam KK, Verosloff MS et al (2020) Cell-free biosensors for rapid detection of water contaminants. Nat Biotechnol 38:1451–1459. https://doi.org/10.1038/s41587-020-0571-7
94. Vilkhovoy M, Horvath N, Shih C-H et al (2018) Sequence specific modeling of E. coli cell-free protein synthesis. ACS Synth Biol 7:1844–1857. https://doi.org/10.1021/acssynbio.7b00465
95. Horvath N, Vilkhovoy M, Wayman JA et al (2020) Toward a genome scale sequence specific dynamic model of cell-free protein synthesis in Escherichia coli. Metab Eng Commun 10: e00113. https://doi.org/10.1016/j.mec.2019.e00113
96. Matsuura T, Tanimura N, Hosoda K et al (2017) Reaction dynamics analysis of a reconstituted Escherichia coli protein translation system by computational modeling. Proc Natl Acad Sci 114: E1336–E1344. https://doi.org/10.1073/pnas.1615351114
97. Ma D, Li Y, Wu K et al (2022) Multi-arm RNA junctions encoding molecular logic unconstrained by input sequence for versatile cell-free diagnostics. Nat Biomed Eng 6:298–309. https://doi.org/10.1038/s41551-022-00857-7
98. Copeland CE, Kim J, Copeland PL et al (2022) Characterizing a new fluorescent protein for a low limit of detection sensing in the cell-free system. ACS Synth Biol 11:2800–2810. https://doi.org/10.1021/acssynbio.2c00180